EVOLUTIONARY AND REVOLUTIONARY TECHNOLOGIES FOR MINING

Committee on Technologies for the Mining Industries
National Materials Advisory Board
Board on Earth Sciences and Resources
Committee on Earth Resources
National Research Council

NATIONAL ACADEMY PRESS
Washington, D.C.

NATIONAL ACADEMY PRESS   2101 Constitution Avenue, N.W.   Washington, DC 20418

NOTICE: The project that is the subject of this report was approved by the Governing Board of the National Research Council, whose members are drawn from the councils of the National Academy of Sciences, the National Academy of Engineering, and the Institute of Medicine. The members of the committee responsible for the report were chosen for their special competences and with regard for appropriate balance.

This study was supported by the U.S. Department of Energy, Office of Industrial Technologies, and the National Institute of Occupational Safety and Health, Grant No. DE-AM01-99PO80016. The views and conclusions contained in this document are those of the authors and do not necessarily reflect the views of the Department of Energy or the National Institute of Occupational Safety and Health.

International Standard Book Number: 0-309-07340-5
Library of Congress Control Number: 2001088181

Copies are available for sale from:
National Academy Press
2101 Constitution Avenue, N.W.
Washington, DC 20418
800-624-6242
202-334-3313 (in the Washington metropolitan area)
http://www.nap.edu

Copies are available in limited supply from:
National Materials Advisory Board
2101 Constitution Avenue, N.W.
Washington, DC 20418
202-334-3505
bmaed@nas.edu

Cover: Photograph of open-pit copper mine at Bingham Canyon, Utah. SOURCE: Kennecott Utah Copper Corporation.

Copyright 2002 by the National Academy of Sciences. All rights reserved.
Printed in the United States of America

THE NATIONAL ACADEMIES

National Academy of Sciences
National Academy of Engineering
Institute of Medicine
National Research Council

The **National Academy of Sciences** is a private, nonprofit, self-perpetuating society of distinguished scholars engaged in scientific and engineering research, dedicated to the furtherance of science and technology and to their use for the general welfare. Upon the authority of the charter granted to it by the Congress in 1863, the Academy has a mandate that requires it to advise the federal government on scientific and technical matters. Dr. Bruce M. Alberts is president of the National Academy of Sciences.

The **National Academy of Engineering** was established in 1964, under the charter of the National Academy of Sciences, as a parallel organization of outstanding engineers. It is autonomous in its administration and in the selection of its members, sharing with the National Academy of Sciences the responsibility for advising the federal government. The National Academy of Engineering also sponsors engineering programs aimed at meeting national needs, encourages education and research, and recognizes the superior achievements of engineers. Dr. Wm. A. Wulf is president of the National Academy of Engineering.

The **Institute of Medicine** was established in 1970 by the National Academy of Sciences to secure the services of eminent members of appropriate professions in the examination of policy matters pertaining to the health of the public. The Institute acts under the responsibility given to the National Academy of Sciences by its congressional charter to be an adviser to the federal government and, upon its own initiative, to identify issues of medical care, research, and education. Dr. Kenneth I. Shine is president of the Institute of Medicine.

The **National Research Council** was organized by the National Academy of Sciences in 1916 to associate the broad community of science and technology with the Academy's purposes of furthering knowledge and advising the federal government. Functioning in accordance with general policies determined by the Academy, the Council has become the principal operating agency of both the National Academy of Sciences and the National Academy of Engineering in providing services to the government, the public, and the scientific and engineering communities. The Council is administered jointly by both Academies and the Institute of Medicine. Dr. Bruce M. Alberts and Dr. Wm. A. Wulf are chairman and vice chairman, respectively, of the National Research Council.

COMMITTEE ON TECHNOLOGIES FOR THE MINING INDUSTRIES

MILTON H. WARD, *Chair*, Ward Resources, Incorporated, Tucson, Arizona
JONATHAN G. PRICE, *Vice-chair*, Nevada Bureau of Mines and Geology, Reno
ROBERT RAY BEEBE, consultant, Tucson, Arizona
CORALE L. BRIERLEY, Brierley Consultancy LLC, Highlands Ranch, Colorado
LARRY COSTIN, Sandia National Laboroatories, Albuquerque, New Mexico
THOMAS FALKIE, Berwind National Resources Corporation, Philadelphia, Pennsylvania
NORMAN L. GREENWALD, Norm Greenwald Associates, Tucson, Arizona
KENNETH N. HAN, South Dakota School of Mines and Technology, Rapid City
MURRAY HITZMAN, Colorado School of Mines, Golden
GLENN MILLER, University of Nevada, Reno
RAJA V. RAMANI, Pennsylvania State University, University Park
JOHN E. TILTON, Colorado School of Mines, Golden
ROBERT BRUCE TIPPIN, North Carolina State University, Asheville
RONG-YU WAN, Newmont Mining Corporation, Englewood, Colorado

National Research Council Staff

TAMARA L. DICKINSON, Study Director
CUNG VU, Senior Program Officer (through April 2000)
TERI G. THOROWGOOD, Research Associate
JUDITH L. ESTEP, Senior Administrative Assistant

NATIONAL MATERIALS ADVISORY BOARD

EDGAR A. STARKE, JR., *Chair*, University of Virginia, Charlottesville
EDWARD C. DOWLING, Cleveland Cliffs, Incorporated, Cleveland, Ohio
THOMAS EAGAR, Massachusetts Institute of Technology, Cambridge
HAMISH FRASER, Ohio State University, Columbus
ALASTAIR M. GLASS, Lucent Technologies, Murray Hill, New Jersey
MARTIN E. GLICKSMAN, Rensselaer Polytechnic Institute, Troy, New York
JOHN A.S. GREEN, The Aluminum Association, Incorporated, Washington, D.C.
THOMAS S. HARTWICK, TRW, Redwood, Washington
ALLAN JACOBSON, University of Houston, Texas
SYLVIA M. JOHNSON, NASA, Ames Research Center, Moffett Field, California
FRANK E. KARASZ, University of Massachusetts, Amherst
SHEILA F. KIA, General Motors Research and Development Center, Warren, Michigan
HARRY A. LIPSITT, Wright State University, Yellow Spring, Ohio
ALAN G. MILLER, Boeing Commercial Airplane Group, Seattle, Washington
ROBERT C. PFAHL, JR., Motorola, Schaumburg, Illinois
JULIA PHILLIPS, Sandia National Laboratories, Albuquerque, New Mexico
HENRY J. RACK, Clemson University, South Carolina
KENNETH L. REIFSNIDER, Virginia Polytechnic Institute and State University, Blacksburg
T.S. SUDARSHAN, Materials Modification, Incorporated, Fairfax, Virginia
JULIA WEERTMAN, Northwestern University, Evanston, Illinois

National Research Council Staff

ARUL MOZHI, Acting Director
JULIUS CHANG, Senior Staff Officer
DANIEL MORGAN, Senior Staff Officer
SHARON YEUNG, Staff Officer
TERI G. THOROWGOOD, Research Associate
DANA CAINES, Administrative Associate
JANICE PRISCO, Administrative Assistant
PATRICIA WILLIAMS, Administrative Assistant

BOARD ON EARTH SCIENCES AND RESOURCES

RAYMOND JEANLOZ, *Chair*, University of California, Berkeley
JOHN J. AMORUSO, Amoruso Petroleum Company, Houston, Texas
PAUL B. BARTON, JR., U.S. Geological Survey (Emeritus), Reston, Virginia
BARBARA L. DUTROW, Louisiana State University, Baton Rouge
ADAM M. DZIEWONSKI, Harvard University, Cambridge, Massachusetts
RICHARD S. FISKE, Smithsonian Institution, Washington, D.C.
JAMES M. FUNK, Equitable Production Company, Pittsburgh, Pennsylvania
WILLIAM L. GRAF, Arizona State University, Tempe
SUSAN M. KIDWELL, University of Chicago, Illinois
SUSAN KIEFFER, Kieffer and Woo, Incorporated, Palgrave, Ontario
PAMELA LUTTRELL, Independent Consultant, Dallas, Texas
ALEXANDRA NAVROTSKY, University of California at Davis
DIANNE R. NIELSON, Utah Department of Environmental Quality, Salt Lake City
JONATHAN G. PRICE, Nevada Bureau of Mines and Geology, Reno

National Research Council Staff

ANTHONY R. DE SOUZA, Staff Director
TAMARA L. DICKINSON, Senior Program Officer
DAVID A. FEARY, Senior Program Officer
ANNE M. LINN, Senior Program Officer
LISA M. VANDEMARK, Program Officer
JENNIFER T. ESTEP, Administrative Associate
REBECCA E. SHAPACK, Research Assistant
VERNA J. BOWEN, Administrative Assistant

COMMITTEE ON EARTH RESOURCES

SUSAN M. LANDON *Chair,* Thomasson Partner Associates, Denver, Colorado
CORALE L. BRIERLEY, Independent Consultant, Highlands Ranch, Colorado
GRAHAM A. DAVIS, Colorado School of Mines, Golden
P. GEOFFREY FEISS, College of William and Mary, Williamsburg, Virginia
JAMES M. FUNK, Equitable Production Company, Pittsburgh, Pennsylvania
ALLEN L. HAMMOND, World Resources Institute, Washington, D.C.
PAMELA D. LUTTRELL, Mobil, Dallas, Texas
JAMES H. McELFISH, Environmental Law Institute, Washington, D.C.
THOMAS J. O'NEIL, Cleveland-Cliffs, Inc., Ohio
DIANNE R. NIELSON, Utah Department of Environmental Quality, Salt Lake City
JONATHAN G. PRICE, Nevada Bureau of Mines and Geology, Reno
RICHARD J. STEGEMEIER, Unocal Corporation, Brea, California
HUGH P. TAYLOR, JR., California Institute of Technology, Pasadena, California
MILTON H. WARD, Ward Resources, Inc., Tucson, Arizona

National Research Council Staff

TAMARA L. DICKINSON, Senior Program Officer
REBECCA E. SHAPACK, Research Assistant

Acknowledgments

This report has been reviewed by individuals chosen for their diverse perspectives and technical expertise in accordance with procedures approved by the National Research Council's Report Review Committee. The purpose of this independent review is to provide candid and critical comments that will assist the authors and the NRC in making their published report as sound as possible and to ensure that the report meets institutional standards for objectivity, evidence, and responsiveness to the study charge. The review comments and draft manuscript remain confidential to protect the integrity of the deliberative process. We wish to thank the following individuals for their participation in the review of this report: Bobby Brown, CONSOL; Harry Conger, Homestake Mining Company; Ed Dowling, Cleveland-Cliffs Incorporated; Deverle Harris, University of Arizona; Mark La Vier, Newmont Mining Company; Debra Stuthsacker, Consultant; and Milton Wadsworth, University of Utah.

While the individuals listed above have provided many constructive comments and suggestions, responsibility for the final content of this report rests solely with the authoring committee and the NRC. The review of this report was overseen by Donald W. Gentry, PolyMet Mining Corporation. Appointed by the National Research Council, he was responsible for making certain that an independent examination of this report was carried out in accordance with institutional procedures and that all review comments were carefully considered. Responsibility for the final content of this report rests entirely with the authoring committee and the institution.

Finally, the committee gratefully acknowledges the support of the staff of the National Research Council. We particularly thank Dr. Tamara L. Dickinson for keeping the committee focused on our charge and for advice and guidance throughout the process. We also thank Judy Estep for able assistance with logistics, Teri Thorowgood for technical matters, and Carol R. Arenberg for editorial assistance in minimizing the use of technical terms such as "blunging," "crud," and "slimes."

Preface

Minerals are basic to our way of living. Essentially everything we use in modern society is a product of the mining, agriculture, or oil and gas industries. Mining is the process of extracting raw materials from the Earth's crust.[1] In fact, mining contributes much in the way of raw material to the other two industries. Mining is important to the United States, which is both a major producer and a major consumer of mineral commodities.

As a major producer in the world markets of metals and other mined products, the United States is a prime developer of mining technology, and American experts work in mining operations throughout the world. No country is entirely self-sufficient in mineral resources, and not every country has high-grade, large, exceptionally profitable mineral deposits. Mining is a global industry, and technologies are rapidly transferred from one country to another.

Mining in the United States is an industry in transition. Environmental considerations are shifting coal production from the East and Southeast to lower sulfur resources in the West. Industrial-mineral mining is projected to expand, in response to increasing consumer demand coupled with limitations on import competition for low-value, bulk-commodity products. The expansion of metal mining in the United States is likely to be small because of diminishing ore grades, regulatory burdens, and limited access to land (although there are some exceptions), as well as higher grade deposits being developed worldwide, including by U.S. companies. Nevertheless, technology will continue to play a vital role in all sectors of mining, as it has in the past, making the products of mining available to consumers and raising standards of living. Technological advancements have been the key to keeping mineral depletion and mineral prices in balance.

In this period of transition, innovation and development will be more important than ever. The U.S. Department of Energy's Office of Industrial Technology and the National Institute for Occupational Safety and Health requested that the National Research Council provide guidance on possible future technological developments in the mining sector. In response to that request the Committee on Technologies for the Mining Industries, composed of experts from academia, industry, state governments, and the national laboratories, was formed. Committee members have recognized expertise in exploration geology and geophysics; mining practices and processes for coal, minerals, and metals; process engineering; resource economics; the environmental impacts of mining; mineral and metal extraction and processing technologies; and health and safety.

The report has identified research areas for new technologies that would address exploration, mining and processing and associated health and safety, and environmental issues. The report calls for enhanced cooperation between government, industry, and academia in mineral research and development, which will be vital for the development of new technologies. The federal government's role is especially important. As Dr. Charles M. Vest, president of MIT, stated when he received the 2000 Arthur M. Bueche Award from the National Academy of Engineering, "The role of the federal government in supporting research and advanced education will remain absolutely essential."

---

[1] As used in this report, the raw materials that are mined include metals, industrial minerals, coal, and uranium, the latter two being raw materials for the production of energy. Liquid and gaseous raw materials from the earth, such as oil and natural gas, are not included, although in-situ mining, which is treated in this report, has several technologies in common with conventional oil and gas recovery.

Contents

FIGURES, TABLES, AND SIDEBARS   xv

EXECUTIVE SUMMARY   1

1   INTRODUCTION   7
      Study and Report, 7

2   OVERVIEW OF TECHNOLOGY AND MINING   10
      Importance of Mining, 10
      Mining and the U.S. Economy, 10
      Overview of Current Technologies, 15
      Industries of the Future Program, 17
      Benefits of Research and Development, 17

3   TECHNOLOGIES IN EXPLORATION, MINING, AND PROCESSING   19
      Introduction, 19
      Exploration, 19
      Mining, 24
      In-situ Mining, 33
      Processing, 37

4   HEALTH AND SAFETY RISKS AND BENEFITS   47
      Size of Equipment, 50
      Automation, 50
      Ergonomics, 51
      Alternative Power Sources, 51
      Noise, 51
      Communications, 51
      Training Technology, 51
      Recommendations, 51

5   RESEARCH OPPORTUNITIES IN ENVIRONMENTAL TECHNOLOGIES   53
      Introduction, 53
      Research Opportunities and Technology Areas, 53
      Recommendations, 58

6		CURRENT ACTIVITIES IN FEDERAL AGENCIES	61
		U.S. Department of Agriculture, 61	
		U.S. Department of Commerce, 61	
		U.S. Department of Energy, 61	
		U.S. Department of Defense, 63	
		U.S. Department of Health and Human Services, 64	
		U.S. Department of the Interior, 64	
		U.S. Department of Labor, 64	
		U.S. Department of Transportation, 64	
		U.S. Environmental Protection Agency, 65	
		National Aeronautics and Space Administration, 65	
		National Science Foundation, 65	
		Nonfederal Programs, 65	
		Recommendations, 65	
7		GOVERNMENT-SPONSORED RESEARCH AND DEVELOPMENT IN MINING TECHNOLOGY	67
		Benefits of Research and Development, 67	
		Role of Government, 67	
		Research and Development in Mining Technology, 68	
		Recommendations, 69	
8		SUMMARY OF CONCLUSIONS AND RECOMMENDATIONS	70
		Importance of Mining to the U.S. Economy, 70	
		Technologies in Exploration, Mining, and Processing, 70	
		Health and Safety Risks and Benefits, 70	
		Research Opportunities in Environmental Technologies, 71	
		Role of the Federal Government, 71	
		Available Research and Technology Resources, 72	

REFERENCES — 74

APPENDIXES
A BIOGRAPHIES OF COMMITTEE MEMBERS — 79
B PRESENTATIONS TO THE COMMITTEE — 82
C AGENCY WEB ADDRESSES — 83

ACRONYMS — 85

Figures, Tables, and Sidebars

FIGURES

2-1a   Major base and ferrous metal producing areas, 13
2-1b   Major precious metal producing areas, 13
2-2a   Major industrial rock and mineral producing areas–Part I, 14
2-2b   Major industrial rock and mineral producing areas–Part II, 14
2-3   Coal-bearing areas of the United States, 16

3-1   Helicopter-borne, aeromagnetic survey system, 22
3-2   Helicopter-borne, aeromagnetic survey system, 22
3-3   Photo of open-pit copper mine at Bingham Canyon, 25
3-4   Photograph of a quarry, 25
3-5   A conceptual representation of the general layout of a modern mine, the methods of mining, and the technology used, 26
3-6   Sample layout of an underground mine, identifying various mining operations and terms, 27
3-7   Photograph of longwall coal mining, 28
3-8   The design of an in-situ well field in Highland Mine, Wyoming, 35

4-1   U.S. mine fatalities, 1910 to 1999, 48
4-2   Nonfatal lost workdays, 1978 to 1997, 48
4-3   U.S. fatality rates, 1931 to 1999, 49
4-4   Nonfatal days-lost rates, 1978 to 1999, 49
4-5   Average dust concentrations for U.S. longwall and continuous mining operations, 50

5-1   Photograph of pit lake, 56

TABLES

ES-1   Key Research and Development Needs for the Mining Industries, 3

1-1   Research Agenda for the Mining Industry, 8
2-1   U.S. Net Imports of Selected Nonfuel Mineral Materials, 11
2-2   U.S. Consumption and Production of Selected Mineral Commodities, 12

3-1   Opportunities for Research and Technology Development in Exploration, 24
3-2   Opportunities for Research and Development in Mining, 34

3-3 Opportunities for Research and Technology Development in In-Situ Mining, 36
3-4 Opportunities for Research and Development in Mineral Processing, 46

4-1 Recommendations for Research and Development in Health and Safety, 52

5-1 Opportunities for Research and Technology Development for Environmental Protection, 58

6-1 Estimates of Mining Research and Development Capabilities of the National Laboratories, 62

8-1 Key Research and Development Needs for the Mining Industries, 71

SIDEBARS

3-1 Examples of Environmental and Health Concerns That Should Be Identified During Exploration, 20
3-2 Models for Ore Deposits with Little Environmental Impact, 21
3-3 Need for Research on Fine Particles and Dust, 37

5-1 Phosphogypsum, 54
5-2 Blue Sky Ideas for Research on Environmental Issues, 60

7-1 Benefits of SXEW to Producers and Consumers, 68

8-1 Potential Revolutionary Developments for Mining, 72
8-2 Basic and Applied Research and Development, 72

Executive Summary

The Office of Industrial Technologies (OIT) of the U.S. Department of Energy commissioned the National Research Council (NRC) to undertake a study on required technologies for the Mining Industries of the Future Program to complement information provided to the program by the National Mining Association. Subsequently, the National Institute for Occupational Safety and Health also became a sponsor of this study, and the Statement of Task was expanded to include health and safety.

The NRC formed a multidisciplinary committee of 14 experts (biographical information on committee members is provided in Appendix A) from academia, industry, state governments, and national laboratories. Committee members have recognized expertise in exploration geology and geophysics; mining practices and processes for coal, minerals, and metals; process engineering; resource economics; the environmental impacts of mining; mineral and metal extraction and processing technologies; and health and safety.

The overall objectives of this study are: (a) to review available information on the U.S. mining industry; (b) to identify critical research and development needs related to the exploration, mining, and processing of coal, minerals, and metals; and (c) to examine the federal contribution to research and development in mining processes. Seven specific tasks are outlined below.

1. Review the importance to the U.S. economy (in terms of production and employment) of the mining industries, including the extraction and primary processing of coal, minerals, and metals.
2. Identify research opportunities and technology areas where advances could improve the effectiveness and increase the productivity of exploration.
3. Identify research opportunities and technology areas where advances could improve energy efficiency, increase productivity, and reduce wastes from mining and processing.
4. Review the federal research and technology resources currently available to the U.S. mining industry.
5. Identify potential safety and health risks and benefits of implementing identified new technologies in the mining industries.
6. Identify potential environmental risks and benefits of implementing identified new technologies in the mining industries.
7. Recommend objectives for research and development in mining and processing that are consistent with the goals of the Mining Industry of the Future Program through its government-industry partnership.

To address this charge the committee held six meetings between March and October 2000. These meetings included presentations by and discussions with the sponsors, personnel from other government programs, and representatives of industry and academia. Individuals who provided the committee with oral or written input are identified in Appendix B. As background material, the committee reviewed relevant government documents and materials, pertinent NRC reports, and other technical reports and literature published through October 2000.

This report is intended for multiple audiences: the Office of Industrial Technologies, the National Institute for Occupational Safety and Health, policy makers, scientists, engineers, and industry associations. Chapter 1 provides background material. Chapter 2 provides an overview of the economic importance of mining and the current state of technology (Task 1). Chapter 3 identifies technologies that would benefit major components of the industry in the areas of exploration, mining, and processing (Tasks 2 and 3). Chapters 4 and 5 identify technologies relevant to health and safety and to the environment, respectively (Tasks 5 and 6). Health, safety, and environmental risks and benefits of individual technologies are also interwoven in the discussions in Chapter 3. Chapter 6 describes current activities in federal government agencies that could be applied to the mining sector (Task 4). Chapter 7 discusses the need for federally sponsored research and development in mining technologies. Chapter 8 summarizes the committee's conclusions and recommendations (Task 7).

IMPORTANCE OF MINING TO THE U.S. ECONOMY

Finding. Mining produces three types of mineral commodities—metals, industrial minerals, and fuels—that all countries find essential for maintaining and improving their standards of living. Mining provides critical needs in times of war or national emergency. The United States is both a major consumer and a major producer of mineral commodities, and the U.S. economy could not function without minerals and the products made from them. In states and regions where mining is concentrated this industry plays an important role in the local economy.

TECHNOLOGIES IN EXPLORATION, MINING, AND PROCESSING

Mining involves a full life cycle from exploration through production to closure with provisions for potential postmining land use. The development of new technologies benefits every major component of the mineral industries: exploration, mining (physical extraction of the material from the Earth), processing, associated health and safety issues, and environmental issues. The committee recommends that research and development be focused on technology areas critical for exploration, mining, in-situ mining, processing, health and safety, and environmental protection. These technology areas are listed in Table ES-1 and are summarized below.

Exploration

Modern mineral exploration is largely technology driven. Many mineral discoveries since the 1950s can be attributed to geophysical and geochemical technologies developed by both industry and government. Further research in geological sciences, geophysical and geochemical methods, and drilling technologies could increase the effectiveness and productivity of mineral exploration. Because many of these areas overlap, developments in one area will most likely cross-fertilize research and development in other areas. In addition, many existing technologies in other fields could be adapted for use in mineral exploration.

Technological development, primarily miniaturization in drilling technologies and analytical tools, could dramatically improve the efficiency of exploration, as well as aid in the mining process. At the beginning of the twenty-first century, even as the U.S. mining industry is setting impressive records in underground and surface mine production, productivity, and health and safety in all mining sectors (metal, industrial minerals, and coal), major technological needs have still not been met. Continued government support for spaceborne remote sensing, particularly hyperspectral systems, will be necessary to ensure that this technology is developed to a stage that warrants commercialization. In the field of geological sciences, increasing support of basic science, including support for geological mapping and geochemical research, would provide a significant, though gradual, increase in the effectiveness of mineral exploration. Filling the gaps in fundamental knowledge, including thermodynamic-kinetic data and detailed four-dimensional geological frameworks of ore systems, would aid mineral exploration and development, as well as mining and mineral processing. Focused research on the development of exploration models, particularly for "environmentally friendly" ore deposits, could yield important beneficial results in the short term. If attention were focused on the most important problems, as identified by industry, the effectiveness of research would be greatly increased.

Mining

In simple terms, mining involves breaking apart in-situ materials and hauling the broken materials out of the mine, while ensuring the health and safety of miners and the economic viability of the operation. A relentless search has been under way since the early 1900s for new and innovative mining technologies that would improve health and safety and increase productivity. In recent decades another driver has been a growing awareness of the adverse environmental and ecological impacts of mining.

Although industry currently supports the development of most new geochemical and geophysical technologies, basic research, such as determining the chemistry, biology, and spectral character of soils, would significantly benefit the minerals industry. For example, uncertainty about rock stability and gas and water conditions that will be encountered during underground mining impedes rapid advances and creates health and safety hazards. As mining progresses to greater depths, increases in rock stress require innovative designs to ensure the short-term and long-term stability of the mine structure. Truly continuous mining will require an accelerated search for innovative fragmentation and material-handling systems. Sensing, analyzing, and communicating data and information will become increasingly important. Mining environments present unique challenges to the design and operation of equipment, which must be extremely reliable. Increasing the productive operating time of equipment and mining systems will require innovative maintenance strategies, supported by modern monitoring technologies.

Substantial research and development opportunities could be investigated in support of both surface and underground mining. The entire mining system—rock fracturing, material handling, ground support, equipment utilization, and maintenance—would benefit from research and development in many sectors. However, focus should be primarily in four

TABLE ES-1 Key Research and Development Needs for the Mining Industries

Research and Development Needs	Exploration, Chapter 3[a]	Mining, Chapter 3[a]	In-Situ, Chapter 3[a]	Processing, Chapter 3[a]	Health & Safety, Chapter 4[a]	Environmental Protection, Chapter 5[a]
Basic Research						
Basic chemistry – thermodynamic and kinetic data, electrochemistry	X		X	X		X
Fracture processes – physics of fracturing, mineralogical complexities, etc.		X	X	X		
Geological, geohydrological, geochemical, and environmental models of ore deposits	X	X	X			X
Biomedical, biochemical, and biophysical Sciences	X	X	X	X	X	X
Applied Research						
Characterization – geology (including geologic maps), hydrology, process mineralogy, rock properties, soils, cross-borehole techniques, etc.	X	X	X	X	X	X
Fracture processes – drilling, blasting, excavation, comminution (including rock-fracturing and rubblization techniques for in-situ leaching and borehole mining)	X	X	X	X		
Modeling and visualization – virtual reality for training, engineering systems, fluid flow	X	X	X	X	X	X
Development of new chemical reagents and microbiological agents for mining-related applications (such as flotation, dissolution of minerals, grinding, classification, and dewatering)			X	X		
Biomedical, biochemical, and biophysical sciences			X	X	X	X
Water treatment						X
Closure					X	X
Alternatives to phosphogypsum production and management						X
Technology Development						
Sensors – analytical (chemical and mineralogical; hand-held and down-hole), geophysical (including airplane drones, shallow seismic data, and hyperspectral data), surface features, personal health and safety, etc.	X	X	X	X	X	X
Communications and monitoring		X		X	X	X
Autonomous mining		X			X	
Total resource recovery without environmental impact		X	X	X		X
Fine and ultrafine mineral recovery (including solid-liquid separation, recovery of ultrafine particles, disposal)				X	X	X
In-situ technologies for low-permeability ores (includes some of the technologies under fracture processes as well as directional drilling, drilling efficiencies, casing for greater depths)	X		X	X		
Biomining		X	X	X		
Fracture processes – applications of petroleum and geothermal drilling technologies to mining	X	X	X			

[a]Justification for including these research and development needs is found in the chapters indicated.

key areas: (1) fracture, fragmentation, and cutting with the goal of achieving continuous mining (while conserving overall energy consumption); (2) sensors and sensor systems for mechanical, chemical, and hydrological applications; (3) data processing and visualization methods that produce real-time feedback; and (4) automation and control systems.

In-Situ Mining

In-situ mining is the removal of a mineral deposit without physically extracting the rock. In-situ leaching is a type of in-situ mining in which metals are leached from rocks by aqueous solutions, a hydrometallurgical process. There are many opportunities for research and technology development related to in-situ mining and related approaches to direct extraction. The chief hurdle to using in-situ leaching with more types of mineral deposits is the permeability of the ore body. Technologies that would fracture and rubblize ore so that fluids would preferentially flow through the ore body and dissolve ore-bearing minerals are a high priority. For some commodities, such as phosphate rock and coal, the removal of the entire mass without dissolving specific minerals through bore-hole mining may be a promising approach.

Key environmental and health concerns related to in-situ leaching are bringing potentially toxic elements or lixiviants to the surface or mobilizing them into groundwater. The development of lixiviants and microbiological agents that could selectively dissolve the desired elements and leave the undesired elements in the rock would be extremely beneficial. The closure of in-situ leaching facilities raises additional environmental concerns. Therefore, research that would increase the overall availability and effectiveness of in-situ mining technologies should also include evaluations of how these facilities could be closed without impacting the long-term quality of groundwater.

Processing

Mineral processing encompasses unit processes for sizing, separating and processing minerals, including comminution, sizing, separation, dewatering, and hydrometallurgical or chemical processing. Research and development would benefit mineral processing in the metal, coal, and industrial mineral sectors. Every unit process—comminution (pulverization), physical separation, and hydrometallurgy/chemical processing—could benefit from technological advances, ranging from a better understanding of fundamental principles to the development of new devices and the integration of entire systems.

Because comminution is extremely energy intensive, the industry would significantly profit from technologies that enhance the efficiency of comminution (e.g., new blasting and ore-handling schemes) and selectively liberate and size minerals. Areas for research include fine-particle technologies, from improved production methods for the ultra-fine grinding of minerals to the minimization of fine-particle production in coal preparation, and the monitoring and controlling of properties of fine particles.

Technology needs in physical separation processes are focused mainly on minimizing entrained water in disposable solids, devising improved magnetic and electrical separators, developing better ore-sorting methods, and investigating selective flocculation applications. Although flotation is a well developed technology, the mining industry would benefit from the availability of more versatile and economic flotation reagents, on-stream analyses, and new cell configurations.

The most important transformation of the mineral industry in the next 20 years could be the complete replacement of smelting by the hydrometallurgical processing of base metals. For this to happen, the trend that began with dump and heap leaching coupled with solvent extraction/electrowinning and that was followed by bioleaching and pressure oxidation would have to be accelerated. Future research and development should be focused on innovative reactor designs and materials, sensors, modeling and simulation, high-pressure and biological basics, leaching, and metal-separation reagents.

HEALTH AND SAFETY RISKS AND BENEFITS

Several factors have contributed to improvements in the overall safety conditions in mines. The U.S. Bureau of Mines (whose health and safety function is now partly handled by the National Institute of Occupational Safety and Health since the U.S. Bureau of Mines closure in 1996) and industry have conducted pioneering research on hazards identification and control. Other factors are major improvements in mine design, the passage of stringent health and safety regulations, and the introduction of more productive systems. Although the frequency of major disasters has been reduced, death and disabling injuries caused by machinery, roof falls, and electrical accidents continue to occur, and are a major concern.

On the health front, miners have long been aware of the hazards posed by the gases, dusts, chemicals, and noise encountered in the work environment and in working under conditions of extreme temperatures (hot or cold) and high altitudes. Although progress has been made, occurrences of silicosis, pneumoconiosis (black lung disease), occupational hearing loss, and other health problems have long been associated with and continue to occur in mining operations. Much remains to be accomplished to make the mine environment healthier.

The committee examined the risks and benefits associated with the introduction of new technologies in terms of equipment size, automation, ergonomics, alternate power sources, noise, communications, and training. Relatively new technologies, such as in-situ mining, better designed equipment, and automation, have reduced exposures to traditional hazards. As production and productivity increase with the increasing size of equipment, exposures to health and safety threats are decreased. At the same time, these advancements may introduce new hazards and in some cases may exacerbate known hazards. Developing the knowledge and skills through education and training to recognize and overcome threats to health and safety during both the design and operational stages of a system is critical.

New monitoring and control systems could effectively address issues related to mining equipment and mine system safety. Advances in industrial training technology have immense potential for improving miner training. Most of these advancements could be realized through combinations of sensors, analyses, visualizations, and communication tools that would enable miners to eliminate hazards altogether or enable them to take steps to avoid an emerging hazard.

Finding. Advances in technology have greatly enhanced the health and safety of miners. However, potential health hazards arising from the introduction of new technologies, which may not become evident immediately, must be addressed as soon as they are identified.

RESEARCH OPPORTUNITIES IN ENVIRONMENTAL TECHNOLOGIES

The mining of coal, base and precious metals, and industrial minerals raises several environmental issues. Some are common to all of these sectors; others are specific to one sector, or even to one commodity within a sector. The creation of large-scale surface disturbances, the production of large volumes of waste materials, and exposures of previously buried geologic materials to the effects of oxidation are intrinsic to the mining industry and continue to present complex environmental problems even when the best available practices are conscientiously followed.

Research options that would provide the greatest environmental benefits for the mining industry would focus primarily on protecting surface and groundwater quality. The most urgent needs are for accurate, real-time methods of characterizing the potential of waste materials that generate acid rock drainage and improved techniques for managing these wastes. Research is also needed to further develop and optimize treatment technologies for acid rock drainage, such as biologic reduction, and to address issues associated with the creation of pit lakes. Improved technologies are also necessary for managing nonacidic wastewaters, including the development of effective, low-cost techniques for removing low concentrations of elements, such as selenium, from large volume flows and removing nitrates from wastewater discharges.

Beneficial research could also be focused on techniques to enhance the long-term environmental stability of closed dump and heap-leaching operations and tailings impoundments. Areas for research include the dewatering of phosphate slimes and other slurried mine wastes, as well as the long-term stability of disposal units for these wastes. Better techniques of recovering methane from underground coal mines would provide significant environmental, health, safety, and economic benefits. Research on technologies to control the emission of fine particulates is also needed.

Finding. The need for a better understanding of the scientific underpinnings of environmental issues and for technologies to address them effectively cannot be overemphasized.

Recommendation. Technologies that attempt to predict, prevent, mitigate, or treat environmental problems will be increasingly important to the economic viability of the mining industry. Improved environmental technologies related to mine closures present the greatest opportunity for increasing productivity and saving energy. Research is also needed on water quality issues related to mine closures, which are often challenging and costly to address for all types of mining.

ROLE OF THE FEDERAL GOVERNMENT

Successful research and development has led to new technologies that have reduced production costs; enhanced the quality of existing mineral commodities; reduced adverse environmental, health, and safety impacts; and created or made available entirely new mineral commodities. Consumers, producers, and the economies of neighboring communities are likely to benefit from the results of further research and development.

Mining companies that would benefit from research and development in exploration, mining, and mineral processing presumably have an incentive to pay for some of the costs. The major concern for public policy, however, is that in commercial firms, areas for research and development are selected based on benefits expected to be captured. The external benefits (i.e., benefits realized by consumers and other producers) of research and development often constitute a large portion of the total benefits.

Government funding for basic research is a dominant factor, and its role in applied research and technology development is significant (NRC, 1995c). Funding for basic research and long-term technology development also leads to benefits for other industries. If funding also involves universities, it can support the training of scientists and engineers (including industry and government professionals, researchers, and trainers of the next generation of employees) who will benefit the mining industry, as well as other technology-intensive sectors of the economy.

Finding. The market will not support an optimal amount of research and development, possibly by a wide margin. Without government support, the private sector tends to underfund research and development, particularly high-risk projects with long-term payoffs.

Finding. Although research in a broad range of fields may eventually have beneficial effects for the mining industry, the committee identified a number of areas in which new basic scientific data or technology would be particularly beneficial (Table ES-1).

Recommendation. The federal government has an appropriate, clear, and necessary role to play in funding research and development on mining technologies. The government should have a particularly strong interest in what is sometimes referred to as high-risk, "far-out," "off-the-path," or "blue-sky" research. A portion of the federal funding for basic research and long-term development should be devoted to achieving revolutionary advances with the potential to provide substantial benefits to both the mining industry and the public.

AVAILABLE FEDERAL RESOURCES

For more than a century the federal government has been involved in research and development for basic industries. In addition, many federal agencies are involved in science, engineering, and technology development that could be useful to the mining industry. Many federal research and development programs dealing with transportation, excavation, basic chemical processes, novel materials, and other subjects could ultimately be beneficial to the mining industry. The only active federal program that deals solely with the development of more efficient and environmentally benign mining technologies is the Mining Industries of the Future Program of the Office of Industrial Technologies of the U.S. Department of Energy.

Finding. The committee recognizes that federal agencies undertake worthwhile research and development for their own purposes. Research and development that could benefit the mining sector of the U.S. economy is being pursued by many federal agencies. The problem is not the lack of skilled researchers but the lack of direct focus on the problems of most interest to the mining industry. It would be helpful if progress in these programs were systematically communicated to all interested parties, including the mining sector.

Recommendation. Because it may be difficult for a single federal agency to coordinate the transfer of research results and technology to the mining sector, a coordinating body or bodies should be established to facilitate the transfer of appropriate, federally funded technology to the mining sector. The Office of Industrial Technologies has made some progress in this regard by organizing a meeting of agencies involved in research that could benefit the mining industry.

Office of Industrial Technology Mining Industries of the Future Program

The OIT has adopted a consortia approach in its Industries of the Future Program, a model that has proved to be extremely successful (NRC, 1997a). The Mining Industries of the Future Program is subject to management and oversight by the U.S. Department of Energy and receives guidance from the National Mining Association and its Technology Committee. The NRC's Committee on Technologies for the Mining Industries recognizes that the research and technology needs of the mining industry draw upon many disciplines, ranging from basic sciences to applied health, safety, and environmental sciences.

Recommendation. Consortia are a preferred way of leveraging expertise and technical inputs to the mining sector, and the consortia approach should be continued wherever appropriate. Advice from experts in diverse fields would be helpful for directing federal investments in research and development for the mining sector. Consortia should include universities, suppliers, national laboratories, any ad hoc groups considered to be helpful, government entities, and the mining industry. The Office of Industrial Technologies should institute periodic, independent program reviews of the Mining Industries of the Future Program to assure that industry needs are being addressed appropriately.

1

Introduction

In 1993, the U.S. Department of Energy (DOE) Office of Industrial Technologies (OIT) designated a group of seven industries as Industries of the Future (IOF). The participating industries were selected because of their high energy use and large waste generation. The original IOF industries included aluminum, chemicals, forest products, glass, metal casting, petroleum refining, and steel. Working through trade associations, OIT asked each industry to provide a vision of its technological future and a road map detailing the research and development required to realize that future. Industry specialists assisted in this process, with industry experts taking the lead in each case.

In 1997, OIT asked the National Research Council (NRC) to provide guidance for OIT's transition to the new IOF strategy. The Committee on Industrial Technology Assessment, formed for this purpose had the specific task of reviewing and evaluating the overall program, reviewing certain OIT-sponsored research projects, and identifying crosscutting technologies (i.e., technologies applicable to more than one industry). The committee focused on three specific areas as examples: intermetallic alloys, manufacturing process controls, and separations technologies. Panels were formed to study each area, and the results were published in separate reports: *Intermetallic Alloy Development: A Program Evaluation* (NRC, 1997a); *Manufacturing Process Controls for the Industries of the Future* (NRC, 1998a); and *Separation Technologies for the Industries of the Future* (NRC, 1999a); and a summary report, *An Evaluation of the Research Program of the Office of Industrial Technologies* (NRC, 1999b). Meanwhile, the IOF program had grown; the agricultural products industry was added in 1996 and the mining industry in 1997.

During the 1990s, the NRC produced several reports focused on the U.S. mining industry. The first and most important of these was *Competitiveness of the U.S. Minerals and Metals Industry,* based on a three-year study commissioned by the U.S. Bureau of Mines (USBM) to assess the global minerals and metals industry; review technologies for use in exploration, mining, minerals processing, and metals extraction; and examine research priorities (NRC, 1990). Although the study did not include coal and industrial minerals, it presented a number of recommendations broadly applicable to the mining industry, the supporting academic community, and the USBM. The report also outlined a research agenda (Table 1-1), which has not been fully achieved.

As a follow-up to that report, USBM in 1993 asked the NRC for an ongoing assessment of the USBM research programs. This assessment was originally intended to be a series of three reports; however, only the reports for 1994 and 1995 were issued because the USBM went out of existence in 1996 (NRC, 1994a, 1995a). These reports document the status of federal institutional capabilities prior to the significant decreases in research that followed the dissolution of the USBM.

Two additional NRC studies are relevant to this report. *Mineral Resources and Society: A Review of the USGS Mineral Resource Surveys Program Plan* is a study of basic and applied research in geology and geophysics (NRC, 1996a). *Hardrock Mining on Federal Lands* included valuable information on environmental impacts and some recommended areas for research (NRC, 1999c).

STUDY AND REPORT

The National Mining Association (NMA) published its vision statement, *The Future Begins with Mining*, in September 1998 (NMA, 1998a) and completed its first roadmap, *Mining Industry Roadmap for Crosscutting Technologies* (NMA, 1998b) shortly thereafter. A second roadmap, *Mineral Processing Technology Roadmap,* was released in September 2000 (NMA, 2000). In 1999, OIT began discussions with the NRC for a study on mining technologies to complement information in the NMA documents. The original statement of task was expanded and a second sponsor, the National Institute for Occupational Safety and Health (NIOSH), was added.

TABLE 1-1 Research Agenda for the Mining Industry

Exploration
- improved spatial and spectral imaging to penetrate foliage and surface cover
- increased digital geophysical coverage of the United States magnetically, gravitationally, radiometrically, and spectrally to 0.5 mile
- improved drilling/sampling techniques and analytical methods to increase basic knowledge

Mining
- geosensing to predict variations in an ore body or coal seam, sense the closeness of geological disturbances, and obtain in-situ measurements of ore grade
- nonexplosive rock fragmentation
- intelligent, cognitive mining systems
- in-situ mining

Mineral Processing
- advances in modeling and automation for computer-controlled operations
- integration of blasting with crushing; use of energy other than electromechanical
- development of more efficient flotation systems

Metal Extraction
- advances in hydrometallurgical and biotechnological processes and reagents

Environmental
- development of a systems approach for environmental issues and waste disposal

SOURCE: NRC, 1990.

The NRC established the Committee on Technologies for the Mining Industries to undertake the study. The committee members, 14 experts from academia, industry, state governments, and the national laboratories, have recognized expertise in exploration geology and geophysics; mining practices and processes for coal, minerals, and metals; process engineering; resource economics; the environmental impacts of mining; mineral and metal extraction and processing technologies; and health and safety. Brief biographies of the committee members are provided in Appendix A.

The overall objectives of this study are: (a) to review available information on the U.S. mining industry; (b) to identify critical needs in research and development related to the exploration, mining, and processing of coal, minerals, and metals; and (c) to examine the federal contribution to research and development in mining processes. The seven specific tasks in the Statement of Task are outlined below:

1. Review the importance to the U.S. economy (in terms of production and employment) of the mining industries, including the extraction and primary processing of coal, minerals, and metals.

2. Identify research opportunities and technology areas in which advances could improve the effectiveness and productivity of exploration.

3. Identify research opportunities and technology areas in which advances could improve energy efficiency and productivity and reduce wastes from mining and processing.

4. Review the federal research and technology resources currently available to the U.S. mining industry.

5. Identify potential safety and health risks and benefits of implementing identified new technologies in the mining industries.

6. Identify potential environmental risks and benefits of implementing identified new technologies in the mining industries.

7. Recommend objectives for research and development in mining and processing that are consistent with the goals of the mining industry of the future through its government-industry partnership.

In this report we do not include downstream processing, such as smelting of mineral concentrates or refining of metals. The discussion is limited to technologies that affect the steps leading to the sale of the first commercial product from extraction. The report does not address broader issues, such as transportation.

To address the charge the committee held six meetings between March and October 2000. The meetings included presentations by and discussions with the sponsors, personnel from other government programs, and representatives of industry and academia. Individuals who provided the committee with oral or written information are identified in Appendix B. As background material, the committee reviewed relevant government documents and materials, pertinent NRC reports, and other technical reports and literature published through October 2000.

Concurrent with the NRC study, NIOSH and OIT commissioned the RAND Science and Technology Policy Institute to conduct a study on critical technologies for mining. The approach adopted for that study involved eliciting a wide range of views through interviews with more than 90 senior personnel (managers and above) from 59 organizations (23 mining companies, 29 service providers, and 7 research/other organizations). Two briefings by representatives of RAND during the course of this study provided preliminary findings on industry trends, mining equipment and processes, and health and safety technologies. However, the RAND report was not available to the committee in time to be used for this study.

This report is intended for multiple audiences. It contains advice for OIT, NIOSH, policy makers, scientists, engineers, and industry associations. Chapter 2 provides an overview of the economic importance of mining and the current state of technology (Task 1). Chapter 3 identifies technologies that would benefit major components of the mining industry in the areas of exploration, mining, and processing (Tasks 2 and 3). Chapters 4 and 5 identify technologies relevant to health and safety and environmental issues, respectively (Tasks 5 and 6). The health, safety, and environmental risks and benefits of individual technologies are also interwoven in the descriptions of individual technologies in Chapter 3. Chapter 6 describes current activities in federal government agencies that could be applicable to the mining sector (Task 4). Chapter 7 discusses the need for federally sponsored research and development in mining technologies. Chapter 8 summarizes the committee's conclusions and recommendations (Task 7).

2

Overview of Technology and Mining

This chapter provides background information on the exploration, mining, and processing of mineral commodities. This is followed by a brief overview of the current state of technology in these fields. The role of research and development in improving technology, and thus in offsetting the adverse effects of mineral-resource depletion over time, are highlighted.

IMPORTANCE OF MINING

Mining is first and foremost a source of mineral commodities that all countries find essential for maintaining and improving their standards of living. Mined materials are needed to construct roads and hospitals, to build automobiles and houses, to make computers and satellites, to generate electricity, and to provide the many other goods and services that consumers enjoy.

In addition, mining is economically important to producing regions and countries. It provides employment, dividends, and taxes that pay for hospitals, schools, and public facilities. The mining industry produces a trained workforce and small businesses that can service communities and may initiate related businesses. Mining also yields foreign exchange and accounts for a significant portion of gross domestic product. Mining fosters a number of associated activities, such as manufacturing of mining equipment, provision of engineering and environmental services, and the development of world-class universities in the fields of geology, mining engineering, and metallurgy. The economic opportunities and wealth generated by mining for many producing countries are substantial.

MINING AND THE U.S. ECONOMY

Mining is particularly important to the U.S. economy because the United States is one of the world's largest consumers of mineral products and one of the world's largest producers. In fact, the United States is the world's largest single consumer of many mineral commodities.

The United States satisfies some of its huge demand for mineral commodities by imports (Table 2-1). For decades, the country has imported alumina and aluminum, iron ore and steel, manganese, tin, copper, and other mineral commodities. Nevertheless, the country is also a major producing country and a net exporter of a several mineral commodities, most notably gold. As Table 2-1 shows, the United States produces huge quantities of coal, iron ore, copper, phosphate rock, and zinc, as well as many other mineral commodities that are either exported directly or used in products that can be exported.

According to the U.S. Geological Survey (USGS), the value of the nonfuel[1] mineral commodities produced in the United States by mining totaled some $39 billion in 1999 (USGS, 2000). The value of processed materials of mineral origin produced in the United States in 1999 was estimated to be $422 billion (USGS, 2000). U.S. production of coal in 1999 was 1.1 billion short tons, which represents an estimated value of $27 billion (EIA, 1999a). However, the true contribution of mining to the U.S. economy is not fully reflected in these figures. For example, the economic impact of energy from coal, which produces 22 percent of the nation's energy and about 56 percent of its electricity, is not included.

The Bureau of Labor Statistics in the U.S. Department of Commerce estimates that the number of people directly employed in metal mining is about 45,000, in coal about 80,000, and in industrial minerals about 114,000 (U.S. Department of Labor, 2000a). Together these figures account for less than 1 percent of the country's total employment in the goods-producing sector (U.S. Department of Labor, 2000a). The low employment number reflects the great advances in technology and productivity in all mining sectors and lower production costs.

---

[1] Does not include coal, uranium, petroleum, or natural gas.

TABLE 2-1 U.S. Net Imports of Selected Nonfuel Mineral Materials

Commodity	Percent	Major Sources (1995-98)[1]
Arsenic trioxide	100	China, Chile, Mexico
Bauxite and alumina	100	Australia, Guinea, Jamaica, Brazil
Bismuth	100	Belgium, Mexico, United Kingdom, China
Columbium (niobium)	100	Brazil, Canada, Germany, Russia
Fluorspar	100	China, South Africa, Mexico
Graphite (natural)	100	Mexico, Canada, China, Madagascar
Manganese	100	South Africa, Gabon, Australia, France
Mica, sheet (natural)	100	India, Belgium, Germany, China
Strontium	100	Mexico, Germany
Thallium	100	Belgium, Mexico, Germany, United Kingdom
Thorium	100	France
Yttrium	100	China, France, United Kingdom, Japan
Gemstones	99	Israel, Belgium, India
Antimony	85	China, Bolivia, Mexico, South Africa
Tin	85	Brazil, Indonesia, Bolivia, China
Tungsten	81	China, Russia, Bolivia, Germany
Chromium	80	South Africa, Russia, Turkey, Zimbabwe
Potash	80	Canada, Russia, Belarus
Tantalum	80	Australia, Thailand, China, Germany
Stone (dimension)	77	Italy, India, Canada, Spain
Titanium concentrates	77	South Africa, Australia, Canada, India
Cobalt	73	Norway, Finland, Canada, Zambia
Rare earths	72	China, France, Japan, United Kingdom
Iodine	68	Chile, Japan, Russia
Barite	67	China, India, Mexico, Morocco
Nickel	63	Canada, Russia, Norway, Australia
Peat	57	Canada
Titanium (sponge)	44	Russia, Japan, Kazakhstan, China
Diamond (dust, grit and powder)	41	Ireland, China, Russia
Magnesium compounds	40	China, Canada, Austria, Greece
Pumice	35	Greece, Turkey, Ecuador, Italy
Aluminum	30	Canada, Russia, Venezuela, Mexico
Silicon	30	Norway, Russia, Brazil, Canada
Zinc	30	Canada, Russia, Peru
Gypsum	29	Canada, Mexico, Spain
Magnesium metal	29	Canada, Russia, China, Israel
Copper	27	Canada, Chile, Mexico
Nitrogen (fixed), ammonia	26	Trinidad and Tobago, Canada, Mexico, Venezuela
Cement	23	Canada, Spain, Venezuela, Greece
Mica, scrap and flake (natural)	23	Canada, India, Finland, Japan
Iron and steel	22	European Union, Canada, Japan, Russia
Lead	20	Canada, Mexico, Peru, Australia
Cadmium	19	Canada, Belgium, Germany, Australia
Iron ore	17	Canada, Brazil, Venezuela, Australia
Sulfur	17	Canada, Mexico, Venezuela
Salt	16	Canada, Chile, Mexico, Bahamas
Silver	14	Mexico, Canada, Peru, Chile
Perlite	13	Greece
Asbestos	7	Canada
Phosphate rock	7	Morocco
Talc	6	China, Canada, Japan
Iron and steel scrap	3	Canada, United Kingdom, Venezuela, Mexico
Beryllium	2	Kazakhstan, Russia, Canada, Germany

[1] In descending order of import share.

Additional mineral commodities for which there are some import dependency include:

Gallium	France, Russia, Canada, Kazakhstan	Rhenium	Chile, Germany, Kazakhstan, Russia
Germanium	Russia, Belgium, China, United Kingdom	Selenium	Canada, Philippines, Belgium, Japan
Indium	Canada, China, Russia, France	Vanadium	South Africa, China
Mercury	Russia, Canada, Kyrgyzstan, Spain	Vermiculite	South Africa, China
Platinum	South Africa, United Kingdom Russia, Germany	Zirconium	South Africa, Australia

SOURCE: USGS, 2000.

TABLE 2-2 U.S. Consumption and Production of Selected Mineral Commodities

	Consumption[a] (percentage of world total)	Production[a] (percentage of world total)
Coal[b]	21	22
Uranium[c]	28	6
Iron ore and steel	14[d]	11[e]
Aluminum and bauxite	33[d]	0[e]
Copper	23[d]	13[e]
Zinc	18[d]	11[e]
Gold	10[d]	15[e]
Phosphate rock	32[d]	30[e]

[a] Consumption is for the processed product (e.g., aluminum and steel); production is for the mined product (e.g., bauxite and ores of uranium, iron, aluminum, copper, and zinc).
[b] EIA, U.S. Department of Energy (*http://www.eia.doe.gov/fuelcoal.html*). Data are totals for anthracite, bituminous coal, and lignite for 1998.
[c] Data are for 1999 (Uranium Institute, 1999).
[d] Calculated based on U.S. consumption data and world production data (USGS, 2000).
[e] Production data from U.S. Geological Survey Mineral Commodity Summaries 2000 (USGS, 2000). Production data are for 1999.

In states and regions where mining is concentrated the industry plays a much more important role in the local economy. Overall, the economy cannot function without minerals and the products made from them. Mining in the United States produces metals, industrial minerals, coal, and uranium. All 50 states mine either sand and gravel or crushed stone for construction aggregate, and the mining of other commodities is widespread. The contribution of mining extends to jobs and related benefits to downstream products, such as automobiles, railroads, buildings, and other community facilities.

Metals

Metal mining, which was once widespread, is now largely concentrated in the West (Figures 2-2a and 2-2b), although it is still important in Michigan, Minnesota, Missouri, New York, and Tennessee. The minerals mined include iron, copper, gold, silver, molybdenum, zinc, and a number of valuable but less common metals. Most are sold as commodities at prices set by exchanges rather than by producers. Moreover, the high value-to-weight ratio of most metals means they can be sold in global markets, forcing domestic producers to compete with foreign operations.

The trend in metal mining has been toward fewer, larger, more efficient facilities. Through mergers and acquisitions, the number of companies has decreased, and foreign ownership has increased. The search for economies of scale has also intensified. Mines now employ fewer people per unit of output, and operators are eager to adopt new technologies to increase their efficiency; which benefits customers and reduces the cost of products. Because metal mines have no control over commodity prices, their prevailing philosophy to survive is that they must cut costs. As a result, most domestic metal mining companies have largely done away with in-house research and development, and many are reluctant to invest in technology development for which there is no immediate need.

Industrial Minerals

Industrial minerals, which are critical raw materials for the construction industry, agriculture, and the chemical and manufacturing sectors of the economy, are produced by more than 6,400 companies from some 11,000 mines, quarries, and plants widely scattered throughout the country (Figure 2-3a and 2-3b). Most industrial minerals have a degree of price flexibility because international competition in the domestic market is limited. Although some companies and plants are large, size is not always necessary for economic success. However, obtaining permits for new mines and quarries is often difficult, especially near urban areas, and this may favor larger operations and more underground mining in the future.

The major industrial materials are crushed stone, sand, and gravel, which are lumped together as "aggregate" and comprise about 75 percent of the total value of all industrial minerals. A wide variety of other materials are also mined, such as limestone, building stone, specialty sand, clay, and gypsum for construction; phosphate rock, potash, and sulfur for agriculture[2]; and salt, lime, soda ash, borates, magnesium compounds, sodium sulfate, rare earth elements, bromine, and iodine for the chemical industries. Industrial materials also include a myriad of substances used in pigments, coatings, fillers and extenders, filtering aids, ceramics, glass, refractory raw materials, and other products.

Certain industrial minerals, such as aggregates and limestone, are sometimes said to have "place value." That is, they are low-value, bulk commodities used in such large quantities that nearby sources are almost mandatory. Competition from imports is generally unlikely, although exceptions can be found. Low production costs combined with low ocean transportation costs allows cement clinker to be imported from Canada, Taiwan, Scandinavia, and China. At one end of the spectrum, some materials, such as domestic high-grade kaolin, require extensive processing and are so valuable that the United States is a major exporter. At the other end, materials such as natural graphite and sheet mica are so rare and domestic sources so poor that the United States imports 100 percent of its needs.

---

[2]Nitrogen, once mined as sodium nitrate, has been extracted from the atmosphere by the Haber ammonia process for nearly a century.

OVERVIEW OF TECHNOLOGY AND MINING 13

FIGURE 2-1a Major base and ferrous metal producing areas. SOURCE: Adapted from USGS, 2000.

FIGURE 2-1b Major precious metal producing areas. SOURCE: Adapted from USGS, 2000.

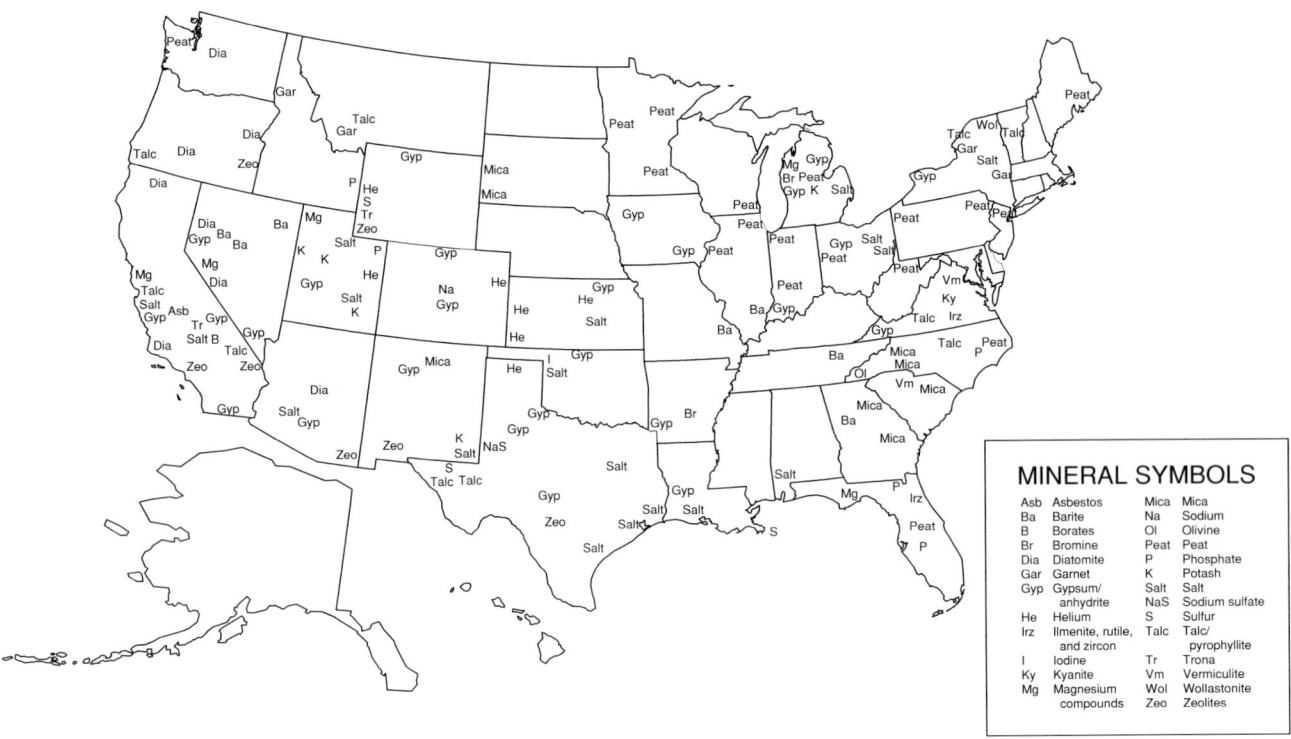

FIGURE 2-2a Major industrial rock and mineral producing areas – Part I. SOURCE: Adapted from USGS, 2000.

FIGURE 2-2b Major industrial rock and mineral producing areas – Part II. SOURCE: Adapted from USGS, 2000.

Unlike the aggregate industry, which is spread over most of the country, some industrial minerals are concentrated in certain parts of the country (Figures 2-3a and 2-3b). Phosphate mining is confined to Florida, North Carolina, Idaho, Utah, and Wyoming. Newly mined sulfur comes from the offshore Gulf of Mexico and western Texas, but recovered sulfur comes from many sources, such as power plants, smelters, and petroleum refineries. The Carolinas and Georgia are the only sources of high-grade kaolin and certain refractory raw materials. The United States has had only one significant rare-earth element mine, located in the desert in southeastern California. Potash, once mined in New Mexico and Utah, now comes mostly from western Canada, where production costs are lower.

The technologies used in the industrial-minerals sector vary widely, from relatively simple mining, crushing, and sizing technologies for common aggregates to highly sophisticated technologies for higher value minerals, such as kaolin and certain refractory raw materials. Agricultural minerals, including phosphates, potash, and sulfur, are in a technological middle range. Uranium can be recovered from phosphate processing. Some investments in new technologies for industrial minerals are intended to increase productivity, but most are intended to produce higher quality products to meet market demands.

Coal and Uranium

Coal is the most important fuel mineral mined in the United States. With annual production in excess of a billion tons since 1994, the United States is the second largest producer of coal in the world. Nearly 90 percent of this production is used for electricity generation; coal accounts for about 56 percent of the electricity generated in the United States (EIA, 1999b). In recent years coal has provided about 22 percent of all of the energy consumed in the United States. Although the nation's reserves of coal are very large, increases in production have been rather small.

Several projections show that coal will lose market share to natural gas, a trend that could be accelerated by concerns over global warming (Abelson, 2000). Coal production may benefit in the short run, however, from electricity deregulation as coal-fired plants use more of their increased generating capacity. With the price of natural gas increasing by more than 100 percent in recent months, projections of future energy mix must be viewed with caution, at least in the short term.

Coal is found in many areas of the United States (Figure 2-4), although there are regional differences in the quantity and quality. Anthracite is found primarily in northeastern Pennsylvania; bituminous coking coals are found throughout the Appalachian region; and other bituminous grades and subbituminous coals are widely distributed throughout Appalachia, the Midwest, and western states.

Deposits of lignite of economic value are found in Montana and the Dakotas, as well as in Texas and Mississippi. Because lignite is about 40 percent water, it is ordinarily used in power plants near the deposits. In recent years considerable research has been focused on making synthetic liquid fuels from lignite.

Some Appalachian and most midcontinent coals have high sulfur contents and thus generate sulfur dioxide when burned in a power plant. Under current environmental regulations effluent gases may have to be scrubbed and the sulfur sequestered. Many power producers have found it more economical to purchase coals from western states. These coals have less sulfur and are preferrable even though they have lower calorific power (energy content). Therefore, the market share of large western mines is increasing. Most western coals are mined from large surface mines, and delivery costs are low because of the availability of rail transportation. Because the capital costs of sulfur scrubbing are high, low-sulfur coal from Montana, Wyoming, and Colorado can be shipped economically by rail over long distances. Concerns about mercury emissions from coal-fired power plants may also influence the future use of coal.

The extensive coal reserves in Utah, Arizona, Colorado, and New Mexico are large enough to produce power to meet local needs, as well as for "wheeling" (transporting energy) over high-voltage transmission lines to Pacific coast states. To serve this market, "mine-mouth" power plants have been built, although air quality and the transmission lines themselves have raised environmental concerns.

Uranium is also mined in the United States. The Energy Information Agency reports that "yellowcake" (an oxide with 91.8 percent uranium) production was 2,300 short tons in 1999 (EIA, 1999d). Overall, nuclear generation produces about 20 percent of the country's electric power (EIA, 1999b). Because the United States is not currently building new nuclear power facilities and because power generation is expanding, uranium's share of electric power generation is likely to fall in the near term. In the longer run, however, the use of uranium in power generation may increase, particularly if the United States seriously attempts to reduce its carbon dioxide emissions. In a recent article in *Science*, Sailor et al. (2000) presented a scenario in which the global carbon dioxide emissions would remain near their present values in 2050, but only by increasing nuclear power generation more than 12-fold.

OVERVIEW OF CURRENT TECHNOLOGIES

The three mining sectors (metals, coal, and industrial minerals) have some common needs for new technologies; other technologies would have narrower applications; and some would be for unique or highly specialized uses. Metal mining can include the following components: exploration and development, drilling, blasting or mechanical excavating, loading, hauling, crushing, grinding, classifying, separating, dewatering, and storage or disposal. Separation may be by physical or

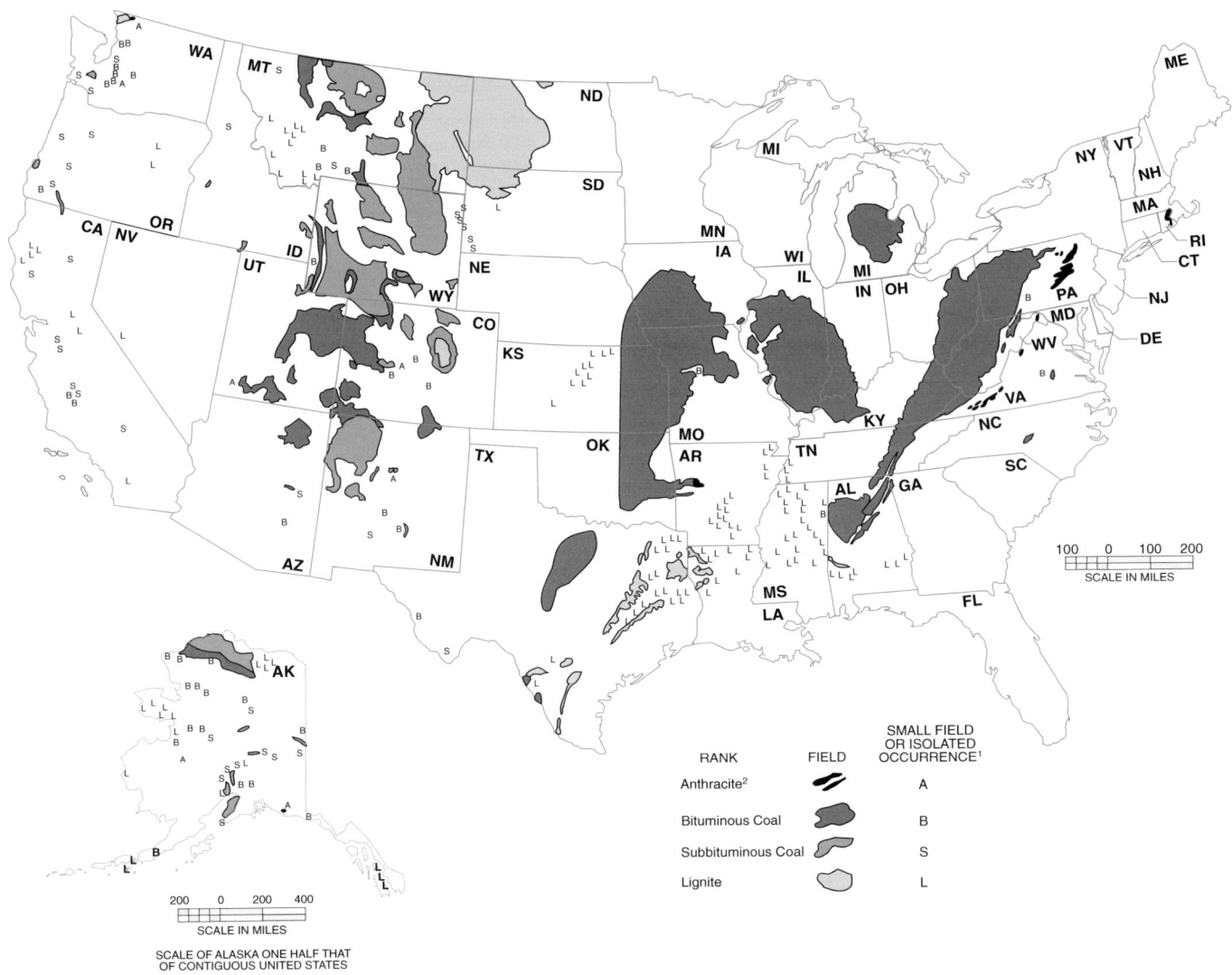

FIGURE 2-3 Coal-bearing areas of the United States. SOURCE: EIA, 1999c.

chemical means, or by a combination of processes; dewatering may be by thickening, filtering, centrifugation, or drying. Storage of metal concentrates may be open or enclosed; disposal of waste products is ordinarily in ponds or dumps. Treatment beyond crushing may be by wet or dry methods; if the latter, dust control is necessary. Classification is usually thought of as discrimination based on size, although with the use of a medium (usually water or air) particles can be differentiated to some degree by mass, or even by shape.

Mining of industrial minerals may include several of the unit operations listed above, but the largest sector of this type of mining (the production of stone, gravel, and sand) seldom requires separation beyond screening, classification, and dense media separation, such as jigging. Other industrial mineral operations require very sophisticated technologies, even by metal-mining standards, to obtain the high quality of certain mineral commodities.

The most common mining methods used by surface coal mines are open pits with shovel-and-truck teams and opencast mines with large draglines. In underground coal mining, the most common methods are mechanical excavation with continuous miners and longwall shearers. Some coals, mostly coals mined underground, may require processing in a preparation plant to produce marketable products. Crushing and screening are common, as are large-scale gravity plants using jigs and dense-media separators, but flotation is not always attractive because of its costs and the moisture content of the shipped product. Coal and coal-bed methane are combustible and sometimes explosive. Therefore, deliberate fine grinding is avoided until just before the coal is burned.

Although the mining industry dates back thousands of years, the industry's technology is quite modern, the result of both incremental improvements and revolutionary developments. Although a miner or explorer, say, 75 years ago might recognize some of the equipment and techniques used today, many important changes have occurred in equipment design and applications. Trucks, shovels, and drills are much larger; electricity and hydraulic drives have replaced compressed air; construction materials are stronger and more durable; equipment may now contain diagnostic computers to anticipate failures; and such equipment usually yields higher productivity, increased margins of safety for workers and the public, and greater environmental protection.

Although incremental improvements have driven much of this progress, major contributions have also come from revolutionary developments. Some examples of revolutionary developments in mining are the use of ammonium-nitrate explosives and aluminized-slurry explosives, millisecond delays in blast ignition, the global positioning system (GPS) in surface-mine operations, rock bolts, multidrill hydraulic jumbos, load-haul-dump units, safety couplers on mine cars, longwall mining, and airborne respirable dust control. In plants there are radiometric density gauges, closed-circuit television, hydrocyclones, wedge-bar screens, autogenous and semiautogenous grinding mills, wrap-around drives, high-intensity magnetic separators, spirals and Reichert cones, high-tension separators, continuous assay systems, high-pressure roll grinding, computerized modeling and process control, and many more innovations. The increase in productivity in the past several decades, made possible by new technologies has far exceeded the average increase for the U.S. economy as a whole.

INDUSTRIES OF THE FUTURE PROGRAM

The goals of the IOF program, namely improving energy efficiency, reducing waste generation, and increasing productivity, present both challenges and opportunities for mining. Exploration normally requires very little energy. However, some exploration techniques, such as satellite remote sensing, require space flights, which use prodigious amounts of energy. Reducing waste generation suggests that more waste be left underground, and this is already being done to a considerable extent in the underground metal-mining sector by returning tailings mixed with cement underground as fill. If in-situ mining is considered as a means of reducing waste, the site-specific nature of this method and its potential environmental effects must be taken into account. Increasing productivity will require increasing output or reducing input, or both.

The IOF program has identified potential areas for improvements in mining. Some enabling tools are already available: sensors, ground-penetrating radar, GPS, and laser measuring techniques. Possible applications in surface and underground mining and milling operations include autonomous robotic equipment, technologies that can "look ahead" of the working face, safer and faster rock bolting closer to the face, and mechanical excavators or drill-blast-load units capable of working close to the face while keeping personnel away from dangerous situations.

Investments in research and development by the mineral industry have been smaller than those of other industries for several reasons. Typically, investment in research and development is risky. Furthermore, the mining industry often considers exploration itself as a form of research. Therefore, rather than investing research funds in the development of new technologies, the industry has invested heavily in exploration to find high-grade, large, or other more attractive deposits, which can lead to better positioning in the competitive business environment.

BENEFITS OF RESEARCH AND DEVELOPMENT

Mineral commodities are extracted from nonrenewable resources, which has raised concerns about their long-term availability. Many believe that, as society exploits its favorable existing mineral deposits and is forced to then exploit poorer quality deposits that are more remote and more difficult to process, the real costs and prices of essential mineral commodities will rise. This could threaten the living standards of future generations and make sustainable development more difficult or impossible. Mineral depletion tends to push up the real prices of mineral commodities over time. However, innovations and new technologies tend to mitigate this upward pressure by making it easier to find new deposits, enabling the exploitation of entirely new types of deposits, and reducing the costs of mining and processing mineral commodities. With innovations and new technologies more abundant resources can be substituted for less abundant resources. In the long run the availability of mineral commodities will depend on the outcome of a race between the cost-increasing effects of depletion and the cost-reducing effects of new technologies and other innovations.

In the past century new technologies have won this race, and the real costs of most mineral commodities, despite their cyclic nature, have fallen substantially (Barnett and Morse, 1963). Real prices, another recognized measure of resource availability, have also fallen for many mineral commodities; although some scholars contend that this favorable trend has recently come to an end (see Krautkraemer [1998] for a survey of the literature in this area). In any case, there is no guarantee that new technologies will keep the threat of mineral depletion at bay indefinitely. However, research and development, along with the new technologies they produce, constitute the best weapon in society's arsenal for doing so.

Mining research and development can lead to new technologies that reduce production costs; it can also enhance the quality of existing mineral commodities while reducing the environmental impacts of mining them and create entirely new mineral commodities. In the twentieth century, for example, the development of nuclear power created a demand for uranium, and the development of semiconductors created a demand for high-purity germanium and silicon.

Another by-product of investment in research and development is its beneficial effect on education. Research funds flowing to universities support students at both the undergraduate and graduate levels and provide opportunities for students to work closely with professors. In a synergistic way research and development funds help ensure that a supply of well-trained scientists and engineers will be available in the future, including individuals who will be working in the fields of exploration, extraction, processing, health and safety, and environmental protection, as well as researchers, educators, and regulators.

The benefits from research and development generally accrue to both consumers and producers, with consumers enjoying most of the benefits over the long run. As both a major consumer and producer of mineral commodities, the United States is particularly likely to benefit greatly from successful research and development in mining technologies.

3

Technologies in Exploration, Mining, and Processing

INTRODUCTION

The life cycle of mining begins with exploration, continues through production, and ends with closure and postmining land use. New technologies can benefit the mining industry and consumers in all stages of this life cycle. This report covers exploration, mining, and processing, but does not include downstream processing, such as smelting of mineral concentrates or refining of metals.

The three major components of mining (exploration, mining, and processing) overlap somewhat. After a mineral deposit has been identified through exploration, the industry must make a considerable investment in development before mining begins. Further exploration near the deposit and further development drilling within the deposit are done while the mining is ongoing. Comminution (i.e., the breaking of rock to facilitate the separation of ore minerals from waste) combines blasting (a unit process of mining) with crushing and grinding (processing steps). In-situ mining, which is treated under a separate heading in this chapter, is a special case that combines aspects of mining and processing but does not require the excavation, comminution, and waste disposal steps. The major components can also be combined innovatively, such as when in-situ leaching of copper is undertaken after conventional mining has rubblized ore in underground block-caving operations.

EXPLORATION

Modern mineral exploration has been driven largely by technology. Many mineral discoveries since the 1950s can be attributed to geophysical and geochemical technologies developed by both industry and government. Even though industrial investment in in-house exploration research and development in the United States decreased during the 1990s, new technologies, such as tomographic imaging (developed by the medical community) and GPS (developed by the defense community), were newly applied to mineral exploration. Research in basic geological sciences, geophysical and geochemical methods, and drilling technologies could improve the effectiveness and productivity of mineral exploration. These fields sometimes overlap, and developments in one area are likely to cross-fertilize research and development in other areas.

Geological Methods

Underlying physical and chemical processes of formation are common to many metallic and nonmetallic ore deposits. A good deal of data is lacking about the processes of ore formation, ranging from how metals are released from source rocks through transport to deposition and post-deposition alteration. Modeling of these processes has been limited by significant gaps in thermodynamic and kinetic data on ore and gangue (waste) minerals, wall-rock minerals, and alteration products. With the exception of proprietary data held by companies, detailed geologic maps and geochronological and petrogenetic data for interpreting geologic structures in and around mining districts and in frontier areas that might have significant mineral deposits are not available. These data are critical to an understanding of the geological history of ore formation. A geologic database would be beneficial not only to the mining industry but also to land-use planners and environmental scientists. In many instances, particularly in arid environments where rocks are exposed, detailed geologic and alteration mapping has been the key factor in the discovery of major copper and gold deposits.

Most metallic ore deposits are formed through the interaction of an aqueous fluid and host rocks. At some point along the fluid flow pathway through the Earth's crust, the fluids encounter changes in physical or chemical conditions that cause the dissolved metals to precipitate. In research on ore deposits, the focus has traditionally been on the location of metal depositions, that is, the ore deposit itself. However, the fluids responsible for the deposit must continue through the crust or into another medium, such as seawater, to maintain a high fluid flux. After formation of a metallic ore deposit, oxidation by meteoric water commonly remobilizes and disperses metals and associated elements, thereby creating geochemical and mineralogical haloes that are used in exploration. In addition, the process of mining commonly exposes ore to more rapid oxidation by meteoric water, which naturally affects the environment. Therefore, understanding the movement of fluids through the Earth, for example, through enhanced hydrologic models, will be critical for future mineral exploration, as well as for effectively closing mines that have completed their life cycle (NRC, 1996b).

The focus of research on geological ore deposits has changed with new mineral discoveries and with swings in commodity prices. Geoscientists have developed numerous models of ore deposits (Cox and Singer, 1992). Models for ore deposits that, when mined, have minimal impacts on the environment (such as deposits with no acid-generating capacity) and for deposits that may be amenable to innovative in-situ extraction will be important for the future. Because the costs of reclamation, closure, postmining land use, and long-term environmental monitoring must be integrated into mine feasibility studies, the health and environmental aspects of an orebody must be well understood during the exploration stage (see Sidebars 3-1 and 3-2). The need for characterizations of potential waste rock and surrounding wall rocks, which may either serve as chemical buffers or provide fluid pathways for escape to the broader environment. Baseline studies to determine hydrologic conditions and natural occurrences of potentially toxic elements in rocks, soils, and waters are also becoming critical. The baseline data will be vital to determining how mining may change hydrologic and geochemical conditions. Baseline climatological, hydrological, and mineralogical data are vital; for example, acid-rock drainage will be greatly minimized in arid climates where natural oxidation has already destroyed acid-generating sulfide minerals or where water flows are negligible.

A wealth of geologic data has been collected for some mining districts, but the data are not currently being used because much of the data is on paper and would be costly to convert to digital format. Individual companies have large databases, but these are not available to the research community or industrial competitors. Ideally, geologic research on ore deposits should be carried out by teams of geoscientists from industry, government, and academia. Industry geoscientists have access to confidential company databases and a focus on solving industrial problems; government and academic geoscientists have access to state-of-the-art analytical tools and a focus on tackling research issues. Currently, geologic research activities in the United States are not well coordinated and are limited primarily to studies of individual deposits by university groups and, to a much lesser extent, by the USGS. More effective research is being carried out in Australia and Canada by industry consortia working with government and academia to identify research problems, develop teams with the skills appropriate to addressing those problems, and pool available funding. Both Canada and Australia have resolved issues of intellectual property rights in the industry-university programs, but these issues have yet to be resolved in the United States.

Geochemical and Geophysical Methods

Surface geochemical prospecting involves analyzing soil, rock, water, vegetation, and vapor (e.g., mercury and hydrocarbons in soil gas) for trace amounts of metals or other elements that may indicate the presence of a buried ore deposit. Geochemical techniques have played a key role in the discovery of numerous mineral deposits, and they continue to be a standard method of exploration. With

---

SIDEBAR 3-1
**Examples of Environmental and Health Concerns
That Should Be Identified During Exploration**

- groundwater and surface water quality
- trace elements in existing soils
- trace elements in ores, particularly elements of concern, such as mercury and arsenic
- the presence of asbestiform minerals associated with industrial-minerals operations
- the potential for acid-rock drainage (amounts of sulfide minerals and buffering minerals, climate, and hydrology)
- location of aquifers in relation to ore bodies
- existence and location of sensitive biological communities
- climatological impacts on mining operations, including precipitation and prevailing winds
- socioeconomic and cultural issues, including sustainable development

> **SIDEBAR 3-2**
> **Models for Ore Deposits with Little Environmental Impact**
>
> Ore has traditionally been defined as natural material that contains a mineral substance of interest and that can be mined at a profit. The costs of mine closure and reclamation of the site now constitute a significant portion of mining cost. Hence, ore bodies that can be mined in a way that produces virtually no waste and that leaves a small surface "footprint" may have distinct economic and environmental advantages over ore bodies that produce large amounts of waste and create large land disturbances. Until recently, these criteria have generally not figured significantly in decisions about mineral exploration. Exploration geologists are now developing new ore-deposit models to improve the chances of finding such "environmentally friendly" ore bodies.
>
> The copper ore bodies mined from 1911 to 1938 at Kennicott, Alaska (now within the Wrangell–St. Elias National Park and Preserve), are examples of potentially environmentally "friendly" ore deposits. The ore bodies consisted of veins of massive chalcocite (a mineral consisting of copper and sulfur). The deposits contained nearly 4.5 million tons of 13 percent copper and 65 grams of silver per ton, some of the highest grade deposits ever mined (Bateman, 1942). The ore at Kennicott contained an amount of copper equivalent to a 100-million-ton typical porphyry copper deposit, which is currently one of the primary types of copper deposits being mined worldwide.
>
> The Kennecott deposits were an economically attractive target for exploration. They were also environmentally attractive because they had a large amount of copper in a small volume of rock, so extraction would cause minimal disturbance, and they consisted primarily of chalcocite with little or no iron sulfide that would produce acid-rock drainage. In addition, their location within massive carbonate rock ensured that any acid generated by the oxidation of sulfides would be quickly neutralized (Eppinger et al., 2000).
>
> Deposits similar to those at Kennicott have not been a target for exploration by many companies primarily because exploration geologists have not developed a robust exploration model for this type of deposit and because their small size makes them difficult to locate. Nevertheless, the development of new, robust models for locating deposits of this and other types of ore bodies that can be mined with little adverse environmental impact could have important economic benefits.

increasingly sophisticated analytical techniques and equipment developed in the past 50 years, exploration geologists have been able to detect smaller and smaller concentrations of the elements of interest. Available analytical tools are sufficient for most types of analyses required by the industry. However, new technologies, such as laser fluorescence scanning and portable X-ray fluorescence, which can directly determine concentrations of elements in rocks, and differential leaching techniques are also being developed and used for exploration. As analytical equipment is miniaturized, inexpensive hand-held devices that could be used in the field or in mines to provide real-time analytical results would significantly benefit both mineral exploration and mining, as well as environmental regulators.

Other research that could benefit the minerals industry includes the development of a more thorough understanding of the media being sampled, such as soils. The complex processes that result in soil formation and the behavior of various elements in different soil types are still poorly understood. A recent NRC report on *Basic Research Opportunities in Earth Sciences* calls for multidisciplinary integrative studies of soils (NRC, 2001). Fundamental research in soil science could produce significant spin-offs that would affect geochemical exploration and would contribute to a more thorough understanding of soil ecology for agriculture. Geoscientists are just beginning to understand how organisms concentrate metals. Even though geobotanical exploration was used by a number of companies in the 1970s and 1980s, research in this field, together with investigations of metal concentrations by other organisms, such as bacteria and fungi, has not been focused on mineral exploration. Other relevant areas of research include soil-gas geochemistry and water geochemistry. The NRC report on *Basic Research Opportunities in Earth Sciences* also highlighted the need for geobiological research (NRC, 2001).

Industrial research and development in geophysical methods of mineral exploration have been ongoing since World War II. Canada has led the world in geophysical innovations, primarily through industry support for academic programs and through in-house corporate development of new techniques. An example of the latter is the recent development by the mining industry of a prototype airborne gravity system. Gravity measurements are a typical means of locating dense metallic mineral deposits and of mapping different rock types in the Earth's crust. However, traditional ground-based surveys are time consuming and therefore expensive. As an NRC report in 1997 pointed out, the ability to gather gravity data from an aircraft would significantly increase productivity and reduce the invasiveness of mineral exploration (NRC, 1997b).

Magnetic surveys are commonly conducted by aircraft that must fly at a fixed distance above the ground surface for optimal data acquisition (Figures 3-1 and 3-2). These surveys are difficult to conduct and risky in rugged terrain. The

FIGURE 3-1 Helicopter-borne, aeromagnetic survey system. SOURCE: Newmont Mining Corporation.

FIGURE 3-2 Helicopter-borne, aeromagnetic survey system. SOURCE: Newmont Mining Corporation.

recent development of drones, primarily by the U.S. military, has made more effective geophysical surveys possible. This technology is currently being explored by industry-government consortia in Australia.

Seismic exploration, although already an integral part of petroleum exploration, is rarely used in mineral exploration. The primary reasons are technological and economic. Current seismic technology is used to gather data at relatively great depths (thousands of meters below those typical of mineral deposits). Near-surface seismic imaging is possible but will require the development of new strategies for collecting and processing the data (NRC, 2000). Typical seismic surveys are expensive in terms of data collection and data processing. New computing capabilities have led to cost reductions although the costs are still beyond most budgets for mineral exploration. Thus, seismic companies have had little financial incentive to engage in this type of research and development, and virtually no governmental support has been available.

Geophysical techniques that can deduce geological structures and changes in physical properties between bore holes, such as cross-bore-hole seismic tomography, are promising technologies (NRC, 1996b).

Remote sensing is the recording of spectral data (visible to infrared and ultraviolet wavelengths) from the Earth's surface via an airborne platform, generally a high-flying aircraft, or from near-Earth orbit (NRC, 2000). Government support was critical in initiating this technology. Current technologies include the Landsat thematic mapper and the enhanced thematic mapper multispectral imager by the United States and high-resolution panchromatic imaging technology (SPOT) developed by the French Space Agency, as well as radar imaging (RadarSat) of topography for cloud-covered or heavily vegetated areas. The U.S. government transferred some existing systems to the commercial sector, and several privately owned satellites are currently in operation and providing detailed (4-meter resolution) multispectral imagery. These data are used by the mineral exploration sector, as well as many other industrial, academic, and government groups. Promising new multispectral technologies are being developed by both government and industry groups. The shuttle radar topographic mapping (SRTM) system will provide high-quality, detailed digital topographic and image data. The advanced spaceborne and thermal emission and reflection (ASTER) mission will provide multiband thermal data.

Hyperspectral technologies are being developed to gather additional data that can be used to map the mineralogy of the ground surface. A high-altitude aircraft system, airborne visible/infrared imaging spectrometer (AVIRIS), has been developed by the National Aeronautics and Space Administration (NASA). Data from this sensor have been successfully used for both mineral exploration and mine closures at several sites in the United States. Spaceborne hyperspectral systems are also being developed. The Hyperion is being readied for deployment on the Earth Observing-1 (EO-1) satellite. Foreign systems include the airborne infrared echelle spectrometer (AIRES) instrument being developed by Australia.

Currently, a number of research challenges are being addressed for hyperspectral technology, especially for spaceborne systems. These include the development of focal planes with adequate signal-to-noise spectral resolution to resolve mineral species of importance and the capability of acquiring data at a 10-meter spatial resolution while maintaining a minimum swath width of 10 kilometers. The focal planes must also be compact, lightweight, have accurate pointing capabilities, and be robust enough to maintain calibration for long-duration spaceflights.

Routine use of existing hyperspectral systems by the minerals industry has been hampered by the unavailability of systems for industrial use, the high cost of hyperspectral data (when available) compared to typical multispectral data, and the need for additional research into the processing of hyperspectral data. Government support for system development and deployment, as well as for basic research on the analysis of hyperspectral data, would ensure that these new technologies would be useful for the mineral exploration industry, as well as for a wide range of other users, including land-use planners and environmental scientists.

Drilling Technologies

Almost all mineral exploration involves drilling to discover what is below the surface. No significant changes in mineral drilling technology or techniques have been made for more than three decades (NRC, 1994b). This contrasts sharply with spectacular advances in drilling technologies, including highly directional drilling, horizontal drilling, and a wide range of drilling tools for the in-situ measurement of rock properties, for the petroleum and geothermal sectors. Mineral exploration involves both percussion and rotary drilling that produce rock chips and intact samples of core. The diameter of mineral exploration drill holes (called slimholes) is generally much smaller than the diameter of either petroleum or geothermal wells. Therefore, many of the down-hole tools used for drilling in the petroleum and geothermal fields are too large to be used in the mineral exploration slimholes. The need for miniaturization of existing drilling equipment is growing not only in the mineral industry but also for NASA to investigate drilling on Mars. The development of guided microdrill systems for the shallow depths of many mineral exploration projects will be challenging.

Drilling generally represents the largest single cost associated with mineral exploration and the delineation of an ore deposit once it has been discovered. Hundreds of drill holes may be required to define the boundaries and evaluate the quality of an orebody. Decreasing the number of drill holes, increasing the drilling rate, or reducing the energy requirements for drilling would have a substantial impact on mineral exploration and development costs. In many situations directional drilling could significantly reduce the number of drill holes required to discover a resource in the ground. Novel drilling technologies, such as down-hole hammers, turbodrills, in-hole drilling motors, and jet drilling systems, have the potential to increase the drilling rate. Novel technologies, together with more efficient rock bits, could also reduce energy requirements for drilling.

Down-hole logging is a standard technique in petroleum exploration. However, it is rarely used in mineral exploration. Standard petroleum well-logging techniques include gamma-ray surveys (to distinguish different rock types based on natural radioactivity), spontaneous potential (to determine the location of shales and zones with saline groundwater), mechanical caliper and dipmeter test (to determine dip and structure of the rock mass penetrated), and a variety of other geophysical tests (resistivity, induction, density, and neutron activation). These tests determine the physical properties of the drilled rock mass and differentiate rock types. Typically, the minerals industry has obtained some of this information by taking samples of rock (either drill chips or drill cores) for analysis. The development of down-hole analytical devices, such as spectrometers, would make it possible to conduct in-situ, real-time analyses of trace elements in the rock mass that could dramatically shorten the time required to determine if a drill hole had "hit" or not. Miniaturization will be necessary for existing down-hole technologies to be used in slimholes.

Drilling and access for drilling generally represent the most invasive aspect of mineral exploration. The environmental impacts of exploration activities could be significantly reduced by the development of drilling technologies that would minimize the footprint of these activities on the ground, such as the miniturization of drilling rigs, the ability to test larger areas from each drill site, and better initial targeting to minimize the number of holes.

Recommendations for Research on Exploration Technologies

Numerous opportunities exist for research and development that would significantly benefit exploration (Table 3-1), many of which involve the application of existing technologies from other fields. Support for technological development, primarily the miniaturization of drilling technologies and analytical tools, could dramatically improve the efficiency of exploration and improve the mining process. Although industry currently supports the development of most new geochemical and geophysical technologies, basic research on the chemistry, biology, and spectral characterization of soils could significantly benefit the mineral industry. Continued government support for spaceborne remote sensing, particularly hyperspectral systems, will be necessary to ensure that this technology reaches a stage at which it could

TABLE 3-1 Opportunities for Research and Technology Development in Exploration

Geologic Methods
- more robust thermodynamic and kinetic geochemical data[a,b]
- new ore-deposit models, particularly for deposits with less environmental impact when mined
- better geohydrological models
- geologic maps of more mineralized areas
- databases for mineralized areas

Geochemical/Geophysical Methods
- hand-held and down-hole analytical instruments[b]
- cross-bore-hole characterization
- better understanding of element mobility in soils
- drones for airborne geophysics
- low-cost, shallow seismic methods
- better interpretation of hyperspectral data

Drilling Technology
- application of existing petroleum and geothermal drilling technologies to minerals sector (directional drilling, better bits, down-hole logging)[b]
- novel drilling techniques (e.g., improvements in slimhole drilling and in-situ measurements)

[a] Highest priority for this category.
[b] Has applications for other aspects of the mining industries.

be commercialized. In the field of geological sciences more support for basic science, including geologic mapping and geochemical research, would provide significant though gradual improvements in mineral exploration. Filling gaps in fundamental knowledge, including thermodynamic-kinetic data and detailed four-dimensional geological frameworks of ore systems, would provide benefits not only for mineral exploration and development but also for mining and mineral processing. The thermodynamic-kinetic data would lead to a better understanding of how the ore systems evolved through time, how the minerals in the ores and waste rocks will react after exposure to postmining changes in hydrology, and how new processing technologies should be developed. The geological framework of an ore system includes the three-dimensional distribution of rock types and structure, such as faults and fractures, as well as the fourth dimension of time—how the rocks and structures formed. This framework is important to successful exploration, efficient mining, and later reclamation. Focused research on the development of exploration models for "environmentally friendly" ore deposits might yield important results in the short term. A mechanism for focusing research on the most important issues, as identified by industry, would help focus industrial, governmental, and academic resources on these problems.

MINING

Mining can be broadly divided into two categories: surface mining and underground mining. Nonentry mining is associated with in-situ mining and augering. Each type of mining has numerous variations, depending on the combination of deposit type, rock strength, depth, thickness, inclination, roof, and floor strata. The extraction of narrow veins, steeply inclined deposits, and deposits at great depth present significant challenges.

Surface mining, wherever applicable, is more advantageous than underground mining in terms of ore recovery, operational flexibility, productivity, safety, and cost. Currently, almost all nonmetallic minerals (more than 95 percent), most metallic ores (more than 90 percent), and a large fraction of coal (more than 60 percent) are mined by surface methods (Hartman, 1987). However, as surface mineral deposits are exhausted, underground mining will inevitably become more prevalent. In addition, as more easily minable deposits are depleted, mining technology and equipment and mining systems for extracting problematic deposits will have to be developed.

Surface Mining

Surface mining is a generic term describing several methods of mining mineral deposits from the surface, which entails removing the vegetation, top soil, and rock (called overburden materials) above the mineral deposit, removing the deposit, and reclaiming the affected land for postmining land use. The most important factors determining whether surface mining can be done today are economic and technical—the price for the product, the cost of production, the quality and quantity of the deposit, the volume of overburden to be removed per ton of the deposit, and the feasibility of reclamation. The practice of surface mining is quite complex and can involve all or several of the following steps: site preparation, overburden drilling and blasting, loading and hauling overburden (waste), drilling and blasting the deposit, loading and hauling the ore, and reclaiming the site.

Surface mining methods can be broadly classified as open-pit mining, which includes quarrying, strip mining, contour mining, dredging, and hydraulic mining. Topography and the physical characteristics of the deposit strongly influence the choice of method. In open-pit mining waste is transported to a disposal site, and the ore is transported to a downstream processing site. This method commonly involves a sequence of benches from the surface to the deposit. As the open pit goes deeper into the ground, all of the benches above are extended outward. In appearance, an open-pit excavation resembles an inverted pyramid with its tip in the Earth (Figure 3-3). Large open-pit copper mines can produce up to a million tons of waste and ore per day and can be mined at that rate for decades. Quarrying is similar to open-pit mining except the term is commonly applied to the extraction of dimension stone and aggregates. Fewer benches are required in quarrying than in open-pit metal mining (Figure 3-4); in quarrying, most of the material extracted is marketable. In area-strip mining a trench is dug

FIGURE 3-3 Photograph of open-pit copper mine at Bingham Canyon. SOURCE: Kennecott Utah Copper Corporation.

FIGURE 3-4 Photograph of a quarry. SOURCE: National Stone, Sand and Gravel Association.

through the overburden to expose the deposit, which is then mined. The trench is then widened by removing the overburden from a parallel adjoining cut and placing it in the previous opening where the deposit has been removed. This method is commonly used in places where the topography and the deposit are generally flat. Reclamation is generally concurrent with mining.

Strip mining is commonly used for mining coal seams and phosphate beds. In hilly terrain the mining of the overburden and the deposit (usually a coal seam) follows the contour around the hill and into the hillside up to the economic limits; hence it is called contour mining. In dredging, a suction device (an agitator and a slurry pump) or other mechanical devices are mounted on a floating barge to dig sand, gravel, or other unconsolidated materials under the water and transport them to land. As the material in a location is exhausted, the dredge moves forward, often constructing and carrying its own lake with it to new ground. Hydraulic mining uses water power to fracture and transport a bench of earth or gravel for further processing. Hydraulic mining is used for placer deposits of gold, tin, and other metals.

Surface mining equipment is similar to construction equipment (e.g., scrapers, bulldozers, drills, shovels, front-end loaders, trucks, cranes, draglines). Surface mining today is characterized by very large equipment (e.g., trucks that can haul more than 300 tons of rock, loading shovels with buckets greater than 36 cubic meters, draglines with buckets greater than 120 cubic meters), and modern technology for planning, designing, monitoring, and controlling operations.

Underground Mining

Underground mining is used when the deposit is too deep for surface mining or there is a restriction on the use of the surface land. The deposit is accessed from the surface by vertical shafts, horizontal adits, or inclines (Figure 3-5). The deposit itself is developed by criss-crossing openings (called levels, cross-cuts, raises, etc.) in the orebody, not only to create blocks of ore to be extracted according to a scheme but also to provide for human access, the transport of ore and waste, and adequate ventilation. The drilling, blasting, loading, and transporting of ore from active working areas (faces) are carried out according to a mining plan. If the deposit is soft, such as coal, potash, or salt, mechanical means can be used to cut and load the deposit, thereby eliminating the need

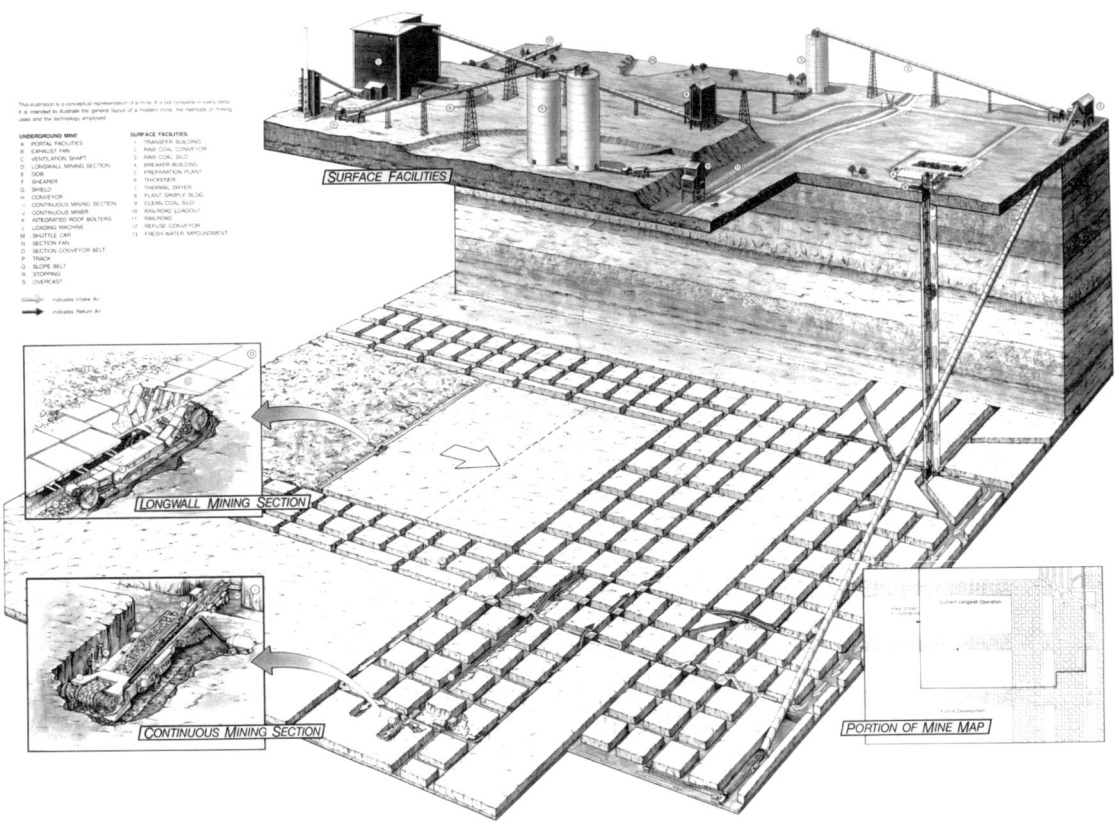

FIGURE 3-5 A conceptual representation of the general layout of a modern mine, the methods of mining, and the technology used. SOURCE: CONSOL, Inc. (now CONSOL Energy, Inc.)

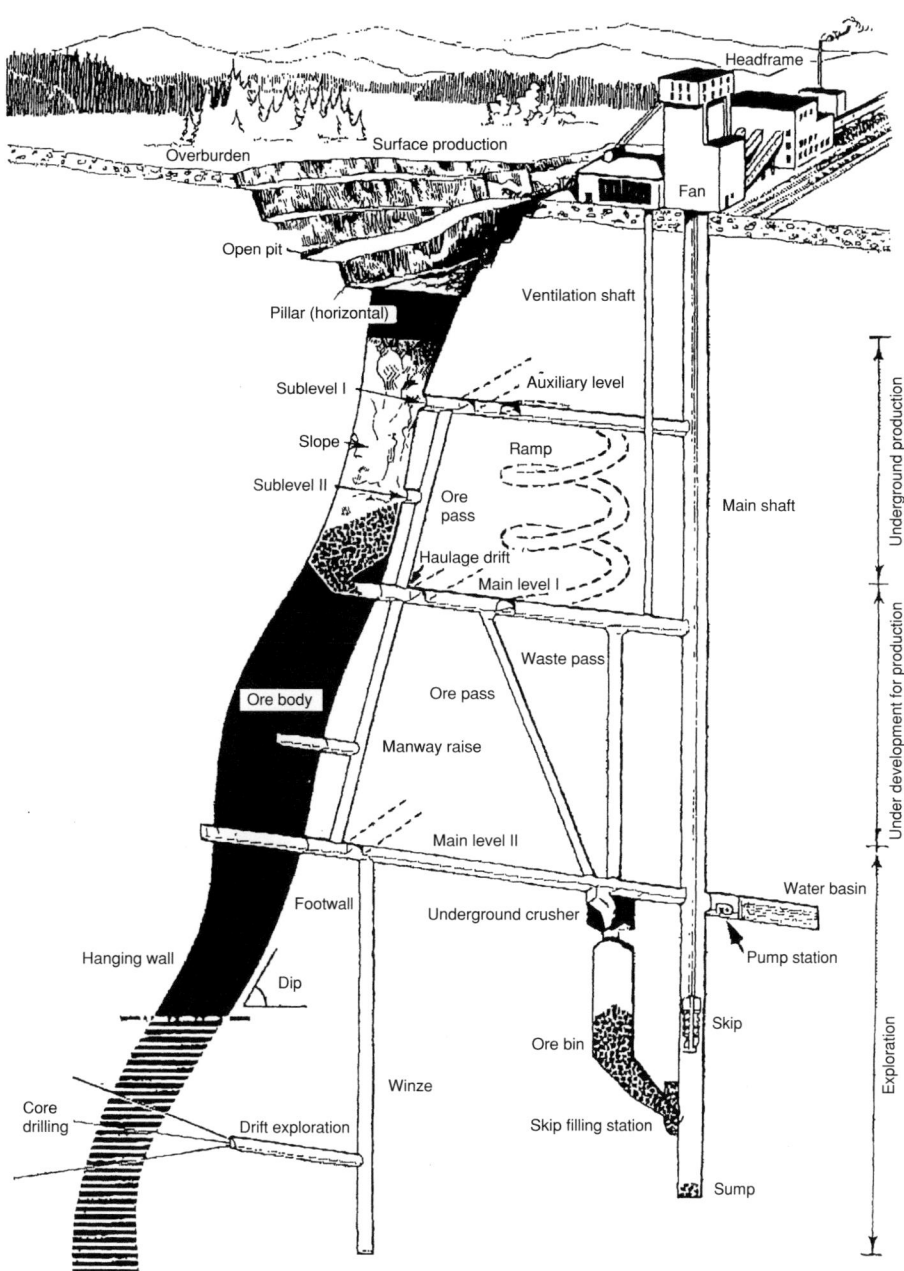

FIGURE 3-6 Sample layout of an underground mine, identifying various mining operations and terms. SOURCE: Hustrulid, 1982.

for drilling and blasting. In hard-rock mines carefully planned drilling into the ore and blasting with dynamite or ammonium-nitrate explosives are common. Underground metal-mining methods may be unsupported, supported, and caving methods, and there are numerous variations of each. Open stopes, room-and-pillar, and sublevel stoping methods are the most common unsupported methods; cut-and-fill stoping when the fill is often waste from the mine and mill tailings is the most common method of supported underground mining (Figure 3-6). Because of the high costs associated with supported and unsupported mining methods, open stoping with caving methods is used whenever feasible.

Underground coal mining today is basically done by two methods: room-and-pillar mining with continuous miners, and longwall mining with shearers. The former is essential for developing large blocks of coal for longwall extraction.

FIGURE 3-7 Photograph of longwall coal mining. SOURCE: CONSOL, Inc. (now CONSOL Energy, Inc.)

The production and productivity of individual, continuous, and longwall production units have increased consistently over the years. In the last two decades longwall mining in the U.S. coal industry has increased from less than 10 percent to nearly 50 percent of the underground tonnage (Fiscor, 1999; NMA, 1999). Currently, about 60 longwall faces produce about 180 million tons of coal per year. However, the production rate depends on the width of the face, the thickness of the seam, and the system for removing the coal from the face.

In longwall mining, operations are concentrated along faces from 250 meters to 350 meters wide. The height of extraction is usually the thickness of the coal seam. The length of the longwall block is about 3,000 meters to 5,000 meters. In a 3-meter-thick coal seam the amount of coal in place in a block is 6 to 7 million tons. The basic equipment is a shearer (a cutting machine) mounted on a steel conveyor that moves it along the face (Figure 3-7). The conveyor discharges the coal onto a conveyor belt for transport out of the mine. The longwall face crew, the shearer, and the face conveyor are under a continuous canopy of steel created by supports called shields. The shields, face conveyor, and shearer are connected to each other and move in a programmed sequence so that the longwall face is always supported as the shearer continuously cuts the coal in slices about 1 meter thick. The shearer is much like a cheese slicer running back and forth across a block of cheese. Modern longwalls are very capital intensive (the equipment alone costs more than $25 million), highly instrumented and automated, employ fewer than six workers at the face, and produce more than 10,000 tons per shift (more than 5 million tons per year).

Technology Needs

In simple terms mining involves breaking in-situ materials and hauling the broken materials out of the mine, while ensuring the health and safety of miners and the economic viability of the operation. Since the early 1900s, a relentless search has been under way for new and innovative mining technologies that can improve health, safety, and productivity. In recent decades another driver has been a growing awareness of the adverse environmental and ecological impacts of mining. Markers along the trail of mining extraction technology include the invention of the safety lamp, and safe use of dynamite for fragmentation, the safe use of electricity, the development of continuous miners for cutting coal, the invention of rock bolts for ground support, open-pit mining technology for mining massive low-grade deposits, the introduction of longwall coal mining, and recently in-situ mining and automated mining.

At the turn of the twenty-first century, even as the U.S. mining industry is setting impressive records in underground and surface mine production, productivity, and health and safety in all sectors of the industry (metal, industrial minerals, and coal), the industry still needs more effective and efficient mining technologies. For example, the inability to ascertain the conditions ahead in the mining face impedes rapid advance and creates health and safety hazards. As mining progresses to greater depths the increase in rock stress requires innovative designs for ensuring the short-term and long-term stability of the mine structure. Truly continuous mining will require innovative fragmentation and material-handling systems. In addition, sensing, analyzing, and communicating data and information will become increasingly important. Mining environments also present unique challenges to the design and operation of equipment. Composed of a large number of complex components, mining systems must be extremely reliable. Therefore, innovative maintenance strategies, supported by modern monitoring technologies, will be necessary for increasing the productive operational time of equipment and the mining system as a whole.

Look-Ahead Technologies

Unexpected geological conditions during the mining process can threaten worker safety and may decrease productivity. Geological problems encountered in mining can include local thinning or thickening of the deposit, the loss of the deposit itself, unexpected dikes and faults, and intersections of gas and water reservoirs. Even with detailed advanced exploration at closely spaced intervals, mining operations have been affected by many problems, such as gas outbursts, water inundations, dangerous strata conditions, and severe operational problems, that can result in injuries to personnel, as well as major losses of equipment and decreases in production. Advances in in-ground geophysics could lead to the development of new technologies for predicting geological conditions in advance of the mining face (defined here as look-ahead technology). Three major technology areas are involved in systems that can interrogate the rock mass ahead of a working face: sensor systems, data processing, and visualization. All three areas should be pursued in parallel to effect progress in the development of a usable system.

Research on the development of specific sensors and sensor systems has focused on seismic methods. In underground mining the mining machine (if mining is continuous) can be used as a sound source, and receivers can be placed in arrays just behind the working face. For drilling and blasting operations, either on the surface or underground, blast pulses can be used to interrogate rock adjacent to the rock being moved. However, numerous difficulties have been encountered, even with this relatively straightforward approach. Current seismic systems are not designed to receive and process multiple signals or continuous-wave sources, such as those from the mining machine.

Some projects sponsored by DOE's Office of Non-Proliferation and National Security have focused on the nature of seismic signals from mining operations to determine whether these signals would interfere with the monitoring of and compliance with the Comprehensive Test Ban Treaty. In another study an NRC panel concluded that controlled blasting methods could generate strong enough signals for analysis and suitable for geotechnical investigations (NRC, 1998b). Other sensing methods that could be explored include electromagnetics and ground-penetrating radar. Combinations of sensing methods should also be explored to maximize the overlaying of multiple data sets.

The second major area that requires additional research is data processing methods for interpreting sensor data. The mining industry has a critical need for processing algorithms that can take advantage of current parallel-processing technologies. Currently, the processing of seismic data can take many hours or days. Real-time turnaround (in minutes) in processing will be necessary for the data to be useful for continuous mining.

The third area of need is data display and visualization, which are closely related to the processing and interpretation of data. The data cannot be quickly assessed unless they are in a form that can be readily reviewed. The need for visualizing data, especially in three dimensions, is not unique to the mining industry. In fact, it is being addressed by many technical communities, especially in numerical analysis and simulation. Ongoing work could be leveraged and extended to meet the needs of the mining industry.

With look-ahead technology unexpected features and events could be detected and avoided or additional engineering measures put in place to prevent injuries and damage to equipment. The economic benefits of anticipating the narrowing or widening of the mined strata or other changes in the geologic nature of the orebody would also be substantial.

Cutting and Fragmentation

Mechanized cutting of rock for underground construction and mining has long been a focus area of technology development (NRC, 1995a). For coal and soft rock, high-production cutting tools and machines have been available for some time and continue to be improved, especially in cutter designs that minimize dust and optimize fragment size for downstream moving and processing. Hardrock presents much more difficult problems. Tunnel-boring machines can cut hardrock at reasonable rates, but the cutters are expensive and wear out rapidly, and the machines require very high thrust and specific energy (the quantity of energy required to excavate a unit of volume). In addition, tunnel-boring machines are not mobile enough to follow sharply changing or dipping ore bodies.

Drilling and blasting methods are commonly used to excavate hardrock in both surface and underground mining. Blasting is also used to move large amounts of overburden (blast casting) in some surface mining operations. Improved blasting methods for more precise rock movement and better control of the fragment sizes would reduce the cost of overbreak removal, as well as the cost of downstream processing.

Recommended areas for research and development in cutting and fragmentation are the development of hardrock cutting methods and tools and improved blast designs. Research on the design of more mobile, rapid, and reliable hardrock excavation would benefit both the mining and underground construction industries. Early focus of this research should be on a better understanding of fracture mechanisms in rock so that better cutters can be designed (NRC, 1996b). In addition, preconditioning the rock with water jets, thermal impulses, explosive impulses, or other techniques are promising technologies for weakening rock, which would make subsequent mechanical cutting easier. Novel combinations of preconditioning and cutting should also be investigated. Numerous ideas for the rapid excavation of hard rock were explored in the early 1970s, motivated by the defense community. These concepts should be re-examined in light of technological improvements in the last 20 years that could make some of the concepts more feasible (Conroy et al., 2000).

Improvements in blast design (e.g., computer-simulation-assisted design) would improve perimeter control, casting, and control of fragment size and would result in large energy savings by decreasing the need for downstream crushing and grinding. New methods of explosive tailoring and timing would also have significant benefits. Research into novel applications of blasting technology for the preparation of in-situ rubble beds for processing would help overcome some of the major barriers to the development of large-scale, in-situ processing methods. New developments in micro-explosives that could be pumped into thin fractures and detonated should be explored for their applications to in-situ fracturing and increasing permeability for processing. These methods would also have applications for coal gasification and in-situ leaching.

The development of better and faster rock-cutting and fragmentation methods, especially for applications to hard rock and in-situ mining, would result in dramatic improvements in productivity and would have some ancillary health and environmental risks and benefits. Mechanized, continuous mining operations are recognized as inherently safer than conventional drill-and-blast mining because it requires fewer unit operations, enables faster installation of ground support, and exposes fewer personnel to hazards. Continuous mining methods for underground hard-rock mining would also raise the level of productivity considerably. The environmental risks associated with in-situ mine-bed preparation by injection of explosives or other means of creating permeability will have to be evaluated. This evaluation should include the hazardous effects of unexploded materials or poisonous by-products in the case of chemical generation of permeability. Current thinking is that these risks would not be high relative to the risks of the processing operations used in in-situ mineral extraction (e.g., retorting and leaching).

Ground Control

The planning and design of virtually all elements of a mining system—openings, roadways, pillars, supports, mining method, sequence of extraction, and equipment—are dictated by the geological and geotechnical characterization of the mine site. The objective of ground control is to use site information and the principles of rock mechanics to engineer mine structures for designed purposes. Massive failures of pillars in underground mines, severe coal and rock bursts, open-pit slope failures, and roof and side falls all represent unexpected failures of the system to meet its design standard. These failures often result in loss of lives, equipment, and in some cases large portions of the reserves. Mining-related environmental problems, such as subsidence, slope instability, and impoundment failures, also reflect the need for more attention to the long-term effects of ground control on mine closures and facility construction.

Advances in numerical modeling, seismic monitoring, acoustic tomography, and rock-mass characterization have contributed immensely to the evolution of modern, ground-control design practices. Problems in mine design and rock engineering are complicated by the difficulties of characterizing rock and rock-mass behavior, inhomogenity and anisotropy, fractures, in-situ stresses, induced stress, and groundwater. The increasing scale of mining operations and equipment, coupled with the greater depths of mining and higher extraction rates, will require improved procedures for ground-control design and monitoring and improved prediction systems for operational ground control.

Site-characterization methods for determining the distributions of intact rock properties and the collective properties of the rock mass will require further development of geostatistical methods and their incorporation into design methodologies for ground support (NRC, 1995b). So far, automated monitoring data, such as data from seismic and/or other geophysical monitoring networks, have not been successfully integrated into the design of mine structures. In addition, ground-support elements, such as rock bolts, could be installed at selected locations and instrumented to monitor stress, support loads, and conditions (to determine maintenance intervals) to validate ground-support designs. With rapid advances in mathematics and numerical modeling, research should focus on approaches, such as real-time analysis and interrogation of data with three-dimensional models. In addition, the heterogeneity of rock strata and the diverse processes acting on the mine system (e.g., geologic, hydrologic, mechanical, and engineering processes) should be considered through stochastic and coupled-system modeling. The technology development advocated for look-ahead technologies should also be beneficial for assessing stability in the immediate vicinity of mining.

The failure of ground control has been a perpetual source of safety and environmental concern. Establishing and adopting better engineering approaches, analytical methods, and design methodologies, along with the other characterization technologies described above, would considerably reduce risks from ground-control failures and provide a safer working environment.

Materials Handling

The design and proper operation of clearance systems for transporting mined materials from the point of mining to processing locations are critical for enhancing production. In many cases the system for loading and hauling the mineral is not truly continuous. Belt and slurry transportation systems have provided continuous haulage in some mining systems. Longwall systems in underground mines, bucket-wheel excavator systems in surface mines, and mobile crushers hooked to conveyor belts in crushed-stone quarries are successful steps in the development of a continuous materials-handling system. Even in these systems haulage is regarded as one of the weakest components. In most cases, both in underground and surface mining, the loading and hauling functions are performed cyclically with loaders and haulers.

The major problem in the development of continuous haulage for underground mining is maneuvering around corners. To increase productivity a truly continuous haulage system will have to advance with the advancing cutter-loader. If the strata conditions require regular support of the roof as mining advances, the support function must also be addressed simultaneously. Therefore, research should also focus on automated roof bolting and integration with the cutting and hauling functions.

The increasing size of loaders and haulers in both surface and underground mines has increased productivity. However, larger equipment is associated with several health and safety hazards from reduced operator visibility. Research should, therefore, focus on advanced technology development for integrating location sensors, obstacle-detection sensors, travel-protection devices, communication tools, and automatic controls.

Reducing the amount of material hauled from underground mines by clearly identifying the waste and ore components at the mine face would result in both energy and cost savings, as well as a reduction in the amount of waste generated. It might even lead to leaving the subgrade material in place through selective mining. For this purpose the development of ore-grade analyzers to quantify the metal and mineral contents in the rock faces would be extremely useful. The ore-grade analyzer must have both real-time analysis and communication capability so operations could be adjusted. Similarly, in surface mines the down-hole analysis of ore in blast holes could lead to more efficient materials handling by identifying ore and waste constituents.

Equally important to improving the performance of materials-handling machinery will be the development of new technologies for monitoring equipment status and for specific automation needs. In addition, for underground applications the interruption of the line of sight with satellites and thus the impossibility of using the GPS means a totally new technology will have to be developed for machine positioning.

Transporting ore for processing can take considerable time and energy and can contribute significantly to the overall cost of production in both surface and underground mining operations. An area for exploratory research should be downstream processing while the ore is being transported. For certain processes transport by conveyer-belt systems and hydraulic transport through pipelines would allow for some processing before the ore reaches the final process mills. Physical separation processes, such as those outlined later in this report, and leaching with certain chemical agents are the most likely processes that could be integrated with transport.

The initial transport of materials is currently done by powered vehicles. In underground mining the use of diesel-powered loading and hauling equipment presents both safety and health challenges. Electric equipment has similar disadvantages, even though it is cleaner and requires less ventilation, because power transmission and cabling for highly mobile equipment complicates operations. Equipment powered from clean, onboard energy sources would alleviate many of these health and safety problems. Research could focus on powering heavy equipment with alternative energy sources, such as new-generation battery technology, compressed air, or novel fuel-cell technology. The development of such technologies may have mixed results from an environmental standpoint. On the one hand, a reduction in the use of fossil fuels would have obvious benefits in terms of reduced atmospheric emissions. On the other hand, the manufacturing and eventual disposal of new types of batteries or fuel could have environmental impacts.

Mining Systems

The industry needs improved overall mining systems. Alternative systems may bear no resemblance to existing systems, although they may be innovative adaptations of the productive components of existing systems (e.g., the deployment of rapid mine-development procedures, truly continuous mining methods, continuous haulage systems, more effective ventilation procedures, and rapid isolation techniques to enhance health, safety, productivity, and resource recovery). From technological and management perspectives several characteristics of a mineral enterprise must be taken into account. Each mineral deposit has unique geological features (e.g., location and physical, mineralogical, chemical characteristics) that have overriding influence on technical and economic decisions. For example, the environment of an underground mine is totally enclosed by surrounding rock. Because mine development is an intensive cash-outflow activity, the current long lead times must be decreased through new technologies.

The problem of low recovery from underground mines is well documented. In underground coal mining the overall recovery in the United States averages about 55 percent; average recovery from longwall mines is about 70 percent (Hartman, 1987). Technology for mining thin coal seams (less than 1 meter thick), particularly thin-seam longwall technology, would be beneficial. In view of the extreme difficulties for workers in such a constricted environment the technology for thin-seam longwalls must include as much automation, remote control, and autonomous operation as possible. Successful longwall and continuous coal mining technology might be adapted to the mining of other laminar-metallic and nonmetallic deposits. Potential problems to be overcome will include the hardness of the ore, the rock conditions and behavior, and the abrasive nature of the mined materials.

Underground mining of thick coal seams (more than 6 meters thick) also presents numerous problems. Current practice is to extract only the best portion of the seam with available equipment. In some cases coal recoveries have been as low as 10 percent. In addition to the sterilization of the resources this practice has created problems of heating and fire. Research should focus on equipment and methods specific to mining thick seams. Hydraulic mining may have potential applications for thick seams. The technical feasibility of hydraulic mining is well established, but equipment and systems that can operate in more diverse conditions will have to be developed. Like the mining of thick coal seams, other mining methods also leave a relatively high percentage of the resource in the ground. Therefore, research could focus on secondary recovery methods (i.e., returning to mined areas to extract resources still in the ground). The petroleum industry has successfully developed secondary recovery methods; steam, carbon dioxide, and water flooding are commonly used to drive oil to the wellheads.

In-situ mining (discussed in more detail later in this chapter) has been remarkably successful for several metallic and nonmetallic deposits. The application of this technique to the secondary recovery of mineral resources is another area for research. Extensive trials on in-situ gasification of coal have been conducted by a number of agencies worldwide, including DOE and the former USBM. In-situ mining has also been attempted for retorting oil shale. The potential benefits of the in-situ gasification of energy resources include reduction of mine development and mining and more efficient use of resources that are otherwise not economical to mine (Avasthi and Singleton, 1983). However, substantial technical problems, including such environmental issues as groundwater contamination, must first be addressed.

A long-standing need of the hardrock mining industry is continuous mining. Currently, only tunnel-boring machines and some prototype road headers have been shown to be capable of mining hardrock. The use of tunnel-boring machines in some mining operations has been limited because they are not very mobile, are difficult to steer, and are completely inflexible in terms of the shape of the mine opening. Tunnel-boring machines are being used more often for mine entry, as in the development of a palladium-platinum mine in Montana. Prototype mobile mining equipment for hardrock was demonstrated in Australia, but production rates were lower than expected, and numerous failures occurred. The solution to this problem will depend largely on the development of advanced cutting technology for hardrock, as well as ways of incorporating new cutting concepts into a mining system that would provide efficient continuous mining with a lower thrust requirement and maximum flexibility. New control systems might incorporate sensor feedback from the cutting head so machine parameters could be adjusted for maximum efficiency. Similar concepts are currently being used in the hydrocarbon drilling industry.

Mining systems that make a clear break with present systems, such as the chemical and biological mining of coal, should also be investigated. In-situ chemical comminution might be possible if the solid coal could be reduced to fragments by treatment with surface-active compounds, such as liquid or gaseous ammonia, and transported to the surface as a suspension in an inert gas. The literature on the biosolubilization of coal and the aerobic and anaerobic conversion of coal by microorganisms and enzymes has been evolving for some time (Catcheside and Ralph, 1997). Biodegradation of coal macromolecules could potentially convert coal carbons to specific, low-molecular-mass products. Research will be necessary to determine the basic mechanisms, as well as to develop conceptual schemes that would make biodegradation cost effective. For all in-situ mining concepts the obvious environmental benefits of limiting surface disturbances and waste generation must be weighed against the potential of adverse impacts on groundwater quality during operation of the mine and upon its closure. Research on chemical or biological mining of coal must also include evaluations of environmental risks posed by reagents and process intermediates.

Improved Machine Performance

Mining depends heavily on mechanical, motor-driven machinery for almost every aspect of the process, from initial extraction to transport to processing. Improving the performance of machinery (thus reducing down time), increasing the efficiency of operation, and lowering maintenance costs would greatly increase productivity. The development and application of better maintenance strategies and more advanced automation methods are two means of improving machine performance.

In recent years new concepts of providing maintenance for large fleets of vehicles, especially vehicles in remote or difficult-to-access areas, have emerged primarily as a result of research sponsored by the U.S. Department of Defense (DOD) and equipment manufacturers. Mining operations are also often conducted in remote locations where access to spare parts and large maintenance facilities may be difficult. Current research has focused on the development of sensor systems that can be incorporated into large vehicles and heavy machinery to monitor continuously the "state of the health" of the vehicle. When problems are detected, the vehicle monitoring system can transmit data directly to a monitoring station at a large repair facility where the problem can be diagnosed, and repair packages can be prepared and shipped to the field before the equipment actually fails. Additional research into sensors, software, and communications could focus on adapting this concept to a variety of mining situations. Leveraging ongoing DOD programs could have substantial payoffs in terms of reduced down time, reduced volume of spare parts stored on site, and lower repair costs.

Better automation and control systems for mining equipment could also lead to large gains in productivity. Some equipment manufacturers are already incorporating human-assisted control systems in newer equipment, and improvements in man-machine interfaces are being made. Additional research should focus on alternatives, however, such as more autonomous vehicles that have both sensor capability and sufficient processing power to accomplish fairly complex tasks without human intervention. Tasks include haulage and mining in areas that are too dangerous for human miners. Semiautonomous control methods should also be explored, such as "fly-by-wire" systems in which the operator's actions do not directly control the vehicle but give directions to a computer, which then decides how to accomplish the action. A good example of this technology is currently being used in large construction cranes; the motion of the crane to move a load from one location to another is controlled by the operator through a computer, which controls the rate of movement of the crane in such a way as to minimize the swing of the load. This technology has considerably improved safety, speeded up cycle time, and enhanced energy conservation in the motion of the crane.

Recommendations

Substantial research and development opportunities could be explored in support of both surface and underground mining. The entire mining system, including rock fracturing, material handling, ground support, equipment utilization, and maintenance, would benefit from research and development in four key areas:

1. fracture, fragmentation, and cutting, with the goal of achieving truly continuous mining in hardrock as is done with coal.
2. small, inexpensive sensors and sensor systems for mechanical, chemical, and hydrological applications.
3. data processing and visualization methods (especially taking advantage of advanced, parallel-computing architecture and methods) that would provide real-time feedback.
4. automation and control systems (especially for mining equipment used in hazardous areas).

The above four areas represent a very broad summary of technology advances that would greatly enhance productivity and safety in mining. A more detailed breakdown is provided in Table 3-2.

IN-SITU MINING

In-situ mining is the "removal of the valuable components of a mineral deposit without physical extraction of the rock" (Bates and Jackson, 1987). In-situ leaching is a type of in-situ mining in which metals or minerals are leached from rocks by aqueous solutions, a hydrometallurgical process (American Geological Institute, 1997). In-situ leaching has been successfully used to extract uranium from permeable sandstones in Texas, Wyoming, and Nebraska, and in-situ leaching of copper has been successfully demonstrated in underground copper mines in Arizona, where prior mining has created sufficient permeability for leaching solutions (lixiviants) to contact ore minerals (Bartlett, 1992, 1998; Coyne and Hiskey, 1989; Schlitt and Hiskey, 1981; Schlitt and Shock, 1979). As used in this report the term in-situ mining includes variations that involve some physical extraction.

In-situ leaching involves the injection of a lixiviant, such as bicarbonate-rich, oxidizing water (with added gaseous oxygen or hydrogen peroxide) in the case of uranium, into the ground to dissolve the metal. The metal is then recovered from the solution pumped to surface-treatment facilities. In-situ leaching technologies are based on geology, geochemistry, solution chemistry, process engineering, chemical engineering, hydrology, rock mechanics and rubblization, and petroleum engineering (Wadsworth, 1983).

Related extraction techniques, herein lumped into the broad category of in-situ mining, include: (1) extraction of water-soluble salts (e.g., halite mined to produce caverns in salt domes in Gulf Coast states); (2) brine extraction (pumping of brines to the surface to remove valuable, naturally dissolved materials, such as lithium in Clayton Valley, Nevada, and zinc from geothermal brines in the Salton Sea in California); (3) sulfur extraction using the Frasch process (wherein hot, high-pressure water is injected to melt sulfur, which is then pumped to the surface, as in west Texas and off-shore Louisiana); (4) bore-hole mining (whereby material is removed by breaking rocks with water jets or other techniques and pumping a slurry of water and broken rock to the surface, an experimental process that was tested on

TABLE 3-2 Opportunities for Research and Development in Mining

Look-Ahead Technologies
- seismic methods and alternatives, such as electromagnetics and ground-penetrating radar
- combinations of sensing methods to provide wider ranges of application and better resolution
- processing algorithms that take advantage of current parallel-processing technologies to provide real-time visualization ahead of the mine face
- visualization of data extended to suit the particular needs of the mining application
- ore-grade analyzers to quantify metal and mineral contents
- down-hole analysis with analyzers for ore and waste interfaces

Cutting and Fragmentation
- cutter designs that optimize fragment size to minimize dust, move materials, and minimize processing
- hardrock cutting methods and tools with lower thrust requirements
- improved blasting methods for better control of fragment sizes and more precise rock movement
- improved explosive tailoring and timing
- blasting technology for the preparation of in-situ rubble beds

Ground Control
- procedures for ground-control design and effective monitoring and prediction systems for operational ground control
- field characterization to determine properties of intact rock and the collective properties of the rock mass
- integration of the automatically monitored data (such as data from a seismic and/or other geophysical data acquisition network) into the design of mine structures
- approaches to facilitate real-time analysis and interrogation of data with 3-D models
- modeling methods that address stochastic features and coupled systems

Materials Handling
- a truly continuous haulage system that advances with the cutter-loader
- automated roof bolting that can be integrated with the cutting and hauling functions
- advanced technology for the integration of location sensors, obstacle-detection sensors, travel-protection devices, automatic controls, and communication tools
- technologies for monitoring the operational status of autonomous operations
- methods of achieving downstream processing while ore is being transported
- alternative energy sources, such as new-generation battery technology, compressed air, and novel fuel-cell technology

Mining Systems
- innovative mine development schemes to reduce lead times and enhance recovery rates
- mining technology—equipment and mining systems—for problematical deposits (e.g., technology for mining thin coal seams, particularly thin-seam longwall mining, and equipment and methods for mining thick coal seams)
- adaptation of longwall and continuous coal mining technology to the mining of other laminar metallic and nonmetallic deposits
- continuous hardrock mining with new cutting concepts incorporated into a continuous mining system
- in-situ gasification of energy resources to address technical problems and environmental issues
- exploration of chemical and biological mining of coal to determine basic mechanisms and develop mining-system concepts
- secondary recovery methods for mining

Improved Machine Performance
- development of sensors, software, and communications for mining situations
- new alternatives for man-machine interfaces
- semiautonomous control methods, such as "fly-by-wire" systems
- more autonomous vehicles that can perform complex tasks without human intervention or oversight

phosphate rock in Florida, uranium-rich sandstones in Wyoming, and bituminous sands in California); and (5) in-situ gasification of coal and in-situ retorting of oil shale (described in the previous section on mining).

In-situ leaching has many environmental advantages over conventional mining because it generates less waste material and causes less surface disturbance (no mill tailings, overburden removal, or waste-rock piles). The major environmental concern is postmining water quality. For example, in the case of uranium, concentrations of uranium and its associated radioactive daughter products and, in some cases, potentially toxic elements, such as arsenic and selenium, could be elevated. Site reclamation has been successful at several south Texas sites where in-situ leaching of uranium was first undertaken in the 1970s. In-situ uranium leaching also has advantages in terms of health and safety because the leaching process selectively removes uranium and leaves most of the dangerous radioactive daughter products in the ground. In addition, little heavy machinery is required to remove the large volumes of rock that would have been processed in a conventional mining operation.

With in-situ leaching low-grade uranium deposits (with approximately 0.1 percent U_3O_8) can be mined; these grades are considerably lower than typical grades in the unconformity-type uranium deposits currently mined in Saskatchewan, Canada (with grades on the order of 4 to 20 percent U_3O_8 [Dennis Stover, vice president, engineering and project development, Rio Algom Mining Corporation, personal communication, June 14, 2000]).

In-situ leaching of uranium typically involves the development of a well field with five-spot injection and production wells (Figure 3-8), four production wells on the corners of a square, and one injection well in the center. Monitor wells, used to monitor fluid flow and containment, are distributed around the periphery of the injection-production well field. Because development of the mine depends heavily on drilling and completion of the well field, improvements in drilling efficiencies (faster and cheaper drilling) would clearly increase the productivity of in-situ mining. With directional drilling, particularly when coupled with sensors on or near the drill bits and controls on water pressures along the length of horizontal segments of holes, lixiviants could be placed more directly in contact with ores (in the middle of the ore bodies).

In-situ leaching of uranium is currently limited to low-grade deposits in highly permeable (hundreds to thousands of millidarcies), essentially horizontal sandstones. Well completions are similar to water wells, with casings perforated in the permeable, ore-bearing aquifers. The use of polyvinyl chloride casing, which is considerably cheaper than steel or stainless steel casing, currently limits depths of economical drilling to within 270 meters of the surface (Dennis Stover, vice president, engineering and project development, Rio Algom Mining Corporation, personal communication, June 14, 2000). The development of inexpensive casing that could withstand higher pressures would expand the resource base to include known deposits at greater depths.

Noninvasive techniques (techniques that do not require drilling holes into the ground) that detect hydrologic inhomogeneities, such as clay lenses that are barriers to fluid flow in sandstones and that determine hydrologic properties (transmissivity, permeability) would greatly improve hydrogeologic modeling and well-field design. Cross-borehole tomography (e.g., three-dimensional seismic or other geophysical imaging similar to magnetic resonance imaging in medicine) is one promising technology. Increased computational speed and greater storage capacity would also improve hydrogeological modeling. Well-field operations can be further improved with the development of in-stream chemical sensors for the major constituents (lixiviants, elements being mined, and elements of environmental concern, such as arsenic, selenium, molybdenum, and vanadium in the case of sandstone uranium deposits).

Thus far in-situ leaching in pristine formations (where the rock matrix has not been modified prior to leaching) has been economically successful only in the highly permeable

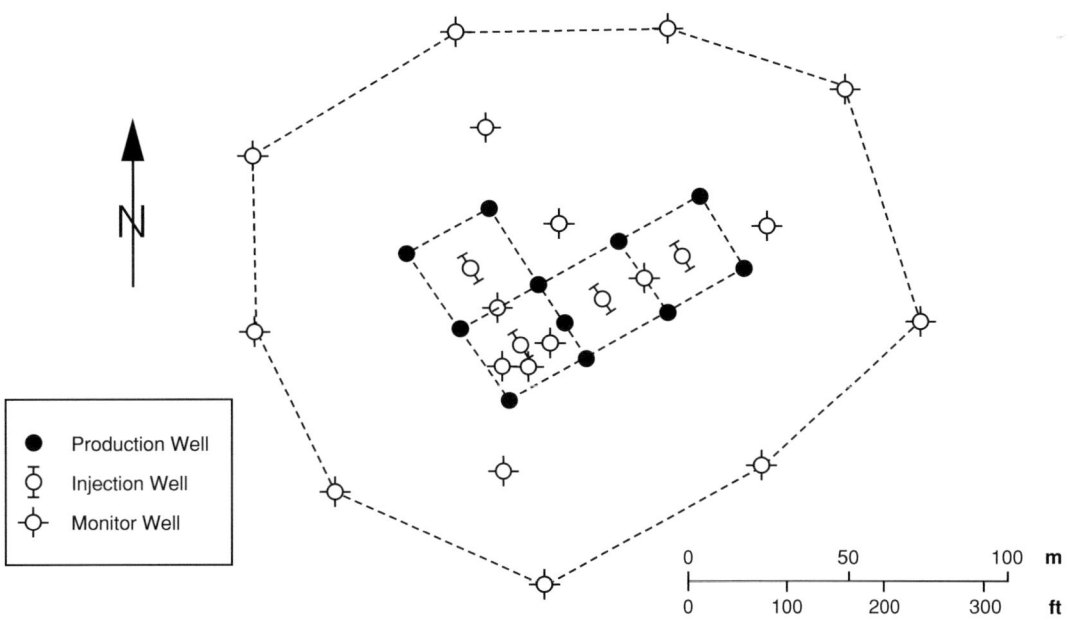

FIGURE 3-8 The design of an in-situ well field in Highland Mine, Wyoming. The drawing shows the locations of wells used to inject lixiviants, wells from which uranium-rich solution is pumped (production wells), and wells used to monitor fluid flow and containment. SOURCE: Adapted from Staub et al., 1986.

sandstones in which some uranium ores are found. Although lixiviants are available to leach various copper oxide and copper sulfide minerals, attempts at in-situ leaching of copper in pristine formations have not been very successful because the lixiviants have not been able to adequately contact the ore minerals in the rock. At the San Manuel in-situ operation in Arizona, recovery rates from caved areas already mined have been on the order of only 50 percent over five years (Sharon Young, consultant, Versitech, Inc., personal communication, June 14, 2000). The most successful in-situ copper leaching has been in ore bodies that had been previously mined; after the high-grade ores were removed open stopes remained with rubble of lower grade wall rock that could be contacted by lixiviants. New technologies for the in-situ fracturing or rubblization of rocks could be extremely beneficial. Increasing permeability in the rocks to allow lixiviants to contact ore minerals is the biggest challenge for the in-situ leaching of metals. One promising approach to increasing permeability, as has been done for copper, is to rubblize rock during conventional mining, thereby taking advantage of the open spaces already created.

Lixiviants are available for leaching not only uranium and copper but also gold, lead, and manganese, to name a few. Nevertheless, cheaper, faster reacting lixiviants would increase production and could also increase the number of metals that could be considered for in-situ leaching. At the same time, lixiviants that suppress the dissolution of undesirable elements, such as arsenic and selenium, which have geochemistries that are significantly different from uranium, would be helpful, as would additives that lower concentrations of those elements during reclamation. Better thermodynamic and kinetic data on important solid phases and aqueous species would facilitate the search for better lixiviants and additives to promote the precipitation or adsorption of undesirable elements. Confinement of lixiviants and mobilized metals to the mining area is another major challenge.

Bore-hole mining has much the same appeal as in-situ leaching because it also tends to minimize the surface footprint of the operation. The biggest challenge for bore-hole mining is the development of tools that can break or cut and remove rock tens of meters beyond the well bores. Various technologies can be envisioned for accomplishing this task; some, such as flexible cutters that can move out from the bore hole in various directions, may require the development of other tools, such as sensors that can distinguish ore from waste rock.

Recommendations for In-Situ Mining Technologies

Many areas offer opportunities for research and technology development in in-situ mining and related approaches to direct extraction (Table 3-3). The chief hurdle to using in-situ leaching for mining more types of mineral deposits is permeability of the ore. The uranium deposits for which in-situ leaching has been successful were located

TABLE 3-3 Opportunities for Research and Technology Development in In-Situ Mining

In-Situ Well-Field Operations
- rock-fracturing and rubblization techniques[a,b]
- directional drilling[b]
- more efficient drilling[b]
- casing for depths below 270 meters
- hydrogeologic modeling[b]
- tomography between bore holes[b]
- sensors for monitoring groundwater and operational controls[b]
- new mining technologies for increasing permeability for in-situ leaching, particularly of base metals

Bore-Hole Excavation
- extending of rock fracturing or cutting to tens of meters beyond well bores[b]
- sensors for assaying samples without removing them[b]

Hydrometallurgical Advances
- development of lixiviants and microbiological agents[a,b]
- suppression of undesirable elements in solution[b]
- additives that precipitate or enhance adsorption of elements of concern during restoration of groundwater quality[b]
- thermodynamic and kinetic data[b]

[a]High priority.
[b]Technologies with applications in other areas of mining and other businesses, such as environmental restoration of metal-contaminated sites.

in exceptionally permeable sandstones. However, ore minerals in the most permeable parts of rock formations are unusual; many metallic ores and industrial-mineral deposits are not highly permeable. Technologies that could fracture and rubblize ore in such a way that fluids would preferentially flow through the orebody and dissolve ore-bearing minerals (although this would be difficult in competent rocks with high compressive strengths) are, therefore, a high priority need for in-situ mining.

For some commodities, such as phosphate rock and coal, removal through bore-hole mining of the entire rock mass without dissolving specific minerals may be an alternate approach. New technologies that would extend rock fracturing and cutting to tens of meters beyond well bores, while maintaining control of the direction of cutting to stay within the orebody or coal seam and avoid removing waste rock, would make bore-hole mining more attractive.

Key environmental and health concerns raised by in-situ leaching are the possibility of potentially toxic elements being brought to the surface or mobilized into groundwater. For example, selenium, arsenic, molybdenum, and radioactive daughter products of uranium are concerns in mining sandstone-type uranium deposits. Therefore, the committee also rates as a high priority development of lixiviants and microbiological agents that can selectively dissolve the desired elements and leave the undesired elements in the rock.

The closure of in-situ leaching facilities raises an additional environmental concern, especially in the copper industry where large-scale in-situ leaching of oxide ore bodies above underground sulfide workings and leaching of sulfide (particularly chalcocite) ores have been conducted. During operations the maintenance of a cone of depression around these ore bodies and the continuous extraction of product solution limits the release of lixiviants and mobilized metals to the surrounding aquifer. However, once mine dewatering and solution recovery are completed, there may be a significant potential for the transport of metals and residual leaching solution. To the extent that the orebody is again totally immersed in the water zone, metals will be in a reduced state, and their mobility will be limited. However, if leaching has taken place above the water table, metals may continue to leach if meteoric water penetration and bacterial activity are sufficient to produce acid conditions. Research should, therefore, also include the evaluation of how these facilities can be closed without long-term adverse impacts to groundwater quality.

PROCESSING

Mineral and coal processing encompasses unit processes required to size, separate, and process minerals for eventual use. Unit processes include comminution (crushing and grinding), sizing (screening or classifying), separation (physical or chemical), dewatering (thickening, filtration, or drying), and hydrometallurgical or chemical processing. Pyrometallurgical processing (smelting of mineral concentrates) is not discussed in this report.

Coal processing, mainly for reducing ash and sulfur contents in the mined raw coal, requires a subset of processing technologies. Some problems in coal processing arise from the way the sulfur and ash are bonded and the need to keep the water content in the cleaned coal low.

Different unit processes described in this section are required in specific cases; some processes are designed especially for the treatment of a particular mineral commodity. Therefore, the committee established a technical framework and broad economic principles as a basis for recommending categories of research and development. The key environmental, health, and safety risks and benefits of these technologies are also highlighted.

Comminution

Comminution, an energy-intensive process, usually begins with blasting of rock in the mining operation followed by crushing in large, heavy machines, often used in stages and in combination with screens to minimize production of particles too fine for subsequent treatment (Sidebar 3-3). Grinding is usually done in tumbling mills, wet or dry, with as little production of fine particles as possible. Comminution is a mature process for which few changes have been made in the past decade. Dry grinding, a higher cost process than wet grinding, is used mainly for downstream processing that requires a dry ground material or for producing a special dry product.

The manner in which rock is blasted in mining operations subjects the rock mass to stress resulting in breakage. Different blasting methods result in different stress distributions in the rock and may have a significant effect on subsequent comminution operations (Chi et al., 1996). The effects of blasting on crushing and grinding are poorly understood. Comminution may take advantage of internal cracking and weakness in the rock caused by an explosive shock from blasting. However, quantifying this phenomenon will require a multidisciplinary investigation involving the physics of rock breakage, mining and mineral processing, and the optimization of energy requirements between blasting and crushing for size reduction.

In the metals and coal industries comminution is generally done to liberate the mineral. In the industrial-mineral sector grinding is more commonly used to meet product specifications or for economic reasons. For example, wet-ground mica commands a much higher price than dry-ground mica of the same quality and size. The grinding method after mineral separation must ensure that the final products

---

SIDEBAR 3-3
Need for Research on Fine Particles and Dust

As processing technologies move toward finer and finer particle sizes, dust and fine particles produced in the mineral industry are becoming an important consideration. Dust is considered dry material; fine particles are suspended in water. The particle sizes of dust and fine particles are defined differently for various sectors of the mineral industry. Unwanted fine particles in the coal industry may be less than 0.147 millimeters (minus 100 mesh particles), while unwanted fine particles for many industrial minerals are less than 10 microns.

Fine particles and dust can represent a health hazard, an environmental concern, and an economic loss. Processes for capturing dust and removing it from the atmosphere, either dry (e.g., in bag houses) or wet (e.g., in scrubbers), are highly efficient. Fine particles are most often disposed of in waste ponds.

The amount of waste dust and fine particles is increasing significantly as more rock is mined and processed. Research should be focused on minimizing the generation of unwanted fine particles and dust or on using these materials as viable by-products.

meet chemical specifications (e.g., iron contamination) or physical specifications (e.g., particle size or shape).

Energy consumption is a major capital and operating cost of mineral beneficiation, and approximately two-thirds of energy processing costs can be attributed to size reduction. Therefore, comminution is often a significant factor in determining economic viability. A savings of a few percent in comminution efficiency may represent a large dollar savings for the overall mining operation.

High-pressure rolls, recently developed in Germany, can significantly reduce specific energy requirements for size reduction (McIvor, 1997). This technology also has downstream processing advantages because it causes microfractures that increase leaching efficiency. High-pressure rolls are currently being used successfully to comminute cement clinker and limestone (McIvor, 1997). The use of high-pressure rolls in the mining industry has been slow, however, because of the high capital cost of the units and because the process has to be dry. Nevertheless, it is evident that mineral liberation could be improved with these devices, and with more experience and research, this technology is expected to gain greater acceptance in metal-processing plants.

About 10 years ago the water-flush crusher attracted renewed interest, and units have been installed in various operating plants in several countries. These crushers operate on a wet slurry-type feed to improve crushing performance and possibly reduce metal wear. The improved water-flush crusher is an example of an incremental improvement of an existing process.

Energy efficiency in size reduction by grinding is typically less than 20 percent, indicating an enormous potential for improvement. Autogenous and semiautogenous mills, which offer economic benefits because of their relatively large scale and simplicity, quickly gained acceptance. These mills are well suited to continuous, high throughput and can be moderately controlled to produce the required distribution of particle size. However, autogenous grinding is only one step in the total comminution process, which includes sizing, pumping, and often crushing. When evaluating a comminution circuit, energy consumption of all aspects of the system should be considered.

The current comminution technology to reduce material to less than 52 microns is inefficient and limited. Relatively few attempts have been made to develop true alternatives to conventional grinding. This represents an excellent opportunity for innovative research that could lead to revolutionary developments that could have dramatic energy savings.

The processing of ultra-fine particles, either occurring naturally in the ore or produced during comminution, is one of the biggest problems facing the mineral industry. Ultra-fine grinding is becoming common for regrinding flotation concentrates and preparing feed for hydrometallurgical processes. Ultra-fine grinding is mandatory in some industries (e.g., mica produced for the paint industry must be ground to below 10 microns). Current ultra-fine grinding by vertical stirred mills has very high energy requirements (Gao et al., 1995; Orumwense and Forssberg, 1992). Energy-efficient ultra-fine grinding devices would be an important contribution for the future of the mineral industry. Some recent grinding installations in Australia have demonstrated potential for ultra-fine grinding with acceptable power consumption (Johnson, 1998). A combination of high-pressure rolls and ultra-fine grinding devices could potentially save energy in the production of ultra-fine particles because they create micro-cracks during the crushing step.

Another emerging technology is optimization and control of component processes of a system that can optimize the energy efficiency of entire operations. Many aspects of optimization and control are mature technologies that are routinely used and are gradually evolving as better sensors and controls become available. Because of the diversity and variability of mineral deposits, process modeling and simulation of total systems in the mining industry is complex and extremely difficult for dynamic in-plant applications. With the advent of high-speed, large-capacity computers, modeling and simulation of individual unit operations have advanced the basic understanding of processes for the industry. Research in this area will be fruitful and should be continued.

The most important objective in comminution is the liberation or breaking apart of desired mineral crystals from unwanted gangue mineral crystals. Effective, reliable analysis of the liberation phenomenon has recently been achieved through imaging analysis and mathematics. Technology has progressed to a point where it is now possible to predict three-dimensional images from two-dimensional analyses in some mineral systems. Refinements in this technology could lead to defining liberation in an ore, thus eliminating overgrinding and reducing both energy usage and excessive loss of fine-grained particles.

The mineral industry needs innovations in instrumentation for size measurements, chemical analysis, and physical characterizations. Instrumentation to measure the physical and chemical properties in core samples, down the bore hole, or in sections of ore at the mine face would enhance subsequent operations by determining the liberation and separation characteristics of minerals before they were processed. With advancing laser technology new instruments may be able to determine the particle-size distribution of fine particles in both aqueous and gaseous suspensions. Flotation is the major concentration process used in the mineral industry, yet there is no good method of characterizing froth quality. Often the instruments are too costly for small and medium-sized operating plants. Although technology in process instrumentation and sensors has significantly advanced in recent years, much still needs to be accomplished.

The end use of most industrial minerals dictates the particle size. Clays, including the important mineral kaolin, occur naturally in fine and ultra-fine sizes, usually not requiring crushing or, at times, even grinding. After grinding to liberate the mineral's quartz, feldspar, and mica for concentration each of the minerals is subjected to another stage of grinding to meet ultra-fine-size specifications for the commercial market, especially as a filler material. No crushing or grinding is required on the ore matrix in Florida phosphates before flotation, but after removal of contaminants the concentrate is ground prior to the production of phosphoric acid. In the aggregate and sand industries a multitude of sized products with different values are routinely produced.

Reducing the cost of energy is one of several factors of interest in the processing of industrial minerals. For fine and ultra-fine grinding, the industry needs better construction materials for equipment because many minerals, such as quartz, are highly abrasive. In recent years some interest has been shown in the development of chemicals called "grinding aids." The results of tests have been mixed, however, and the economic benefits uncertain. Further research on using chemicals to reduce the cost of fine and ultra-fine grinding appears to be warranted.

Coal processors have an urgent need for a comminution system that minimizes the production of fine particles. The treatment of fine coal particles (less than 0.5 millimeter) costs three to four times that of the treatment of coarse coal particles (more than 0.5 millimeter). In addition, the moisture content of fine particles is usually more than four times that of coarse particles, representing an added penalty.

Physical Separation

Physical separation involves (1) the separation of various minerals from one another and (2) the separation of solids (minerals) from liquid (water). The brief discussion that follows includes only the primary processes for mineral separation. Flotation is unquestionably the most important and widely used process to separate minerals, including metals, industrial minerals (Lefond, 1975), and coal.

Almost all separation processes are conducted in a slurry of water. The vast majority of minerals are concentrated by wet processes, but all mineral products are marketed as low-moisture materials. These processes include gravity separation techniques and flotation. Water is one of the most important parameters in wet-separation techniques. Most mineral plants operate in a closed water cycle by regulation because process water often raises environmental concerns (Ripley et al., 1996). Therefore, dewatering is considered an important step in most processes and is a separate topic for research.

Most physical separation processes are conducted wet, but the availability and cost of water are becoming concerns for most mineral-processing operations. A number of physical separations are conducted on dry feeds, often for reasons having to do with the separation process itself. Dry processes include electrostatic and electrodynamic separation, dry magnetic separation, air tabling, air elutriation, dry cycloning, and mechanized sorting. Many industrial-mineral separations are also dry processes. For example, beach-sand processing for titanium, zirconium, rare earth elements, and some radioactive minerals depends on dry-separation methods. Dry-feed separation processes are usually developed or improved by vendors and users, but additional research would be justified.

Gravity Separation

Gravity separation (including processes that use other forces as adjuncts) is not used much in processes for metal ores because sources of ores amenable to gravity separation are now rare. Exceptions include free gold particles because of the great disparity in density between gold and the common gangue minerals, and tin, titanium, zirconium, and certain rare-earth-elements minerals, which can be efficiently concentrated by combinations of gravity, magnetic, and electrical processes. Innovations continue to be made in gravity separation techniques for metallic minerals, as well as for certain industrial-mineral processes, but mature technologies and machine designs are adequate for metal ores and coarse coal. Innovations could be made, however, with the development of inexpensive gravity separation methods that could be used to recover small quantities of heavy minerals from metal-mining flotation tailings. The use of multiforce fields in the separation of particles could improve gravity separation in combination with other processes.

Some gravity separation methods can be used to treat fine particles if there are large density differences between the desired and undesired minerals. In gold plants, for example, a number of gravity devices, old and new, are being used to recover relatively coarse gold. Over the past few years gravity separators that take advantage of differential specific gravities in a high-gradient centrifugal force field (e.g., Knelson and Falcon separators) have been used successfully for gold. Older devices (such as spirals on which the centrifugal forces are lower, pinched sluices, and Reichert cones) have been adapted for other heavy minerals.

Heavy-media or dense-media separation uses a suspension of fine, heavy minerals (magnetite or ferrosilicon) to ensure that the apparent density of the slurry is intermediate between the density of the heavy and light particles. The light particles float to the surface and are separated. Commonly, separation occurs in the settling tank vessel. In some cases a cyclone is used to provide centrifugal force to assist in the mineral separation. The mineral used as media is recycled magnetically. This method is widely used for coal and to remove shale from construction aggregates. Early work has been done to develop a low-cost, effective, safe, and environmentally acceptable "true" heavy fluid but has not led to a commercial success (Khalafalla and Reimers, 1981). Research is still needed on metallurgically efficient, cost-effective technologies for the metal and nonmetal industries.

Most gravity concentrators operate in dilute pulp systems allowing minerals to separate, in part, according to their specific gravity, usually in conjunction with other forces, such as those imparted by flowing water films and centrifugal force. These processes can be used on finer solids if the differences in specific gravity are sufficiently large or if there are marked differences in shape. The natural viscosity of water and the apparent viscosity of the pulp are the dominant process factors. Low pulp-density feed limits the throughput capacity of the machines and results in high water requirements for the system. Improving gravity separation in dense pulps could increase the number of applications for this technology. Research could make a significant and revolutionary change in the use of gravity concentration for fine and ultra-fine mineral separations. At the present time the only large-scale ultra-fine mineral separation process is the degritting of clay using centrifuges.

Magnetic and Electrical Separation

Magnetic separation, which can be either a dry or a wet process, exploits the differences in magnetic susceptibility of minerals. Electrostatic separation is a dry process in which particles falling through a high-voltage static field are diverted according to their natural charges. Electrostatic separation is not suited to extremely fine particles or to large particles whose masses overcome the electrical effect. Electrodynamic separation (high tension) applies a surface charge to fine particles that then contact a grounded roll. Particles that lose their charges are quickly repelled from the roll; others cling to the roll and fall or are brushed off. Eddy-current separators can treat nonmagnetic conductors that, when exposed to an electrical field, experience a force caused by internal eddy currents and are diverted.

Conventional, low-intensity magnetic separators are widely used on ferromagnetic minerals. Electromagnets are being replaced by stronger, more efficient permanent magnets that can be operated wet or dry. At the next level of magnetic intensity, dry separators are common, and wet high-intensity separators are in everyday use on hematite, a paramagnetic mineral. Many attempts have been made to develop continuously operating magnetic separators with superconducting coils, but batch-type separators for removing fine impurities in the production of high-grade kaolin are the only units that have been successfully commercialized. Further research could be focused on continuously operating magnetic separators for minerals.

Sorting

Various sorting systems are used in other industries to separate materials, but few mining companies use them. Ore sorting in mining is usually considered a preconcentration method of upgrading run-of-mine ore before another beneficiation process. Ore sorting is a dry process primarily used for very coarse particles. Minerals in ore can be separated by color, particle shape, particle size, or some other physical characteristic, most commonly optical properties.

A workable ore-sorting system located at the mine could significantly reduce transportation costs, provide a method of maintaining a constant grade of feed to the process plant, and reduce operating costs by preventing uneconomical material from being processed. With technology advancing so rapidly in the instrumentation and electronics industries, sorting methods may improve sufficiently to be useful in mining.

Flotation

Flotation is both a revolutionary unit process and a mature technology that has been used for approximately 100 years for mineral separation throughout the world. The flotation process, which is versatile, can separate minerals as large as 3.3 millimeters (6 mesh) and as small as 5 microns and can handle minerals with a specific gravity as high as 19 (gold) and as low as 1.5 (vermiculite). The process can be used in a medium of almost pure water, seawater or saturated brines. Mineral separations have been made in water near freezing temperature, as well as in water near 38°C (100°F). Operating flotation plants process as little as 100 tons per day to more than 100,000 tons per day. Flotation is a major separation method for metals, coal, and industrial minerals. Most sulfides can be economically recovered by the flotation process.

Parameters that influence flotation can be divided into two general categories: (1) the surface characteristics of the minerals and (2) the design of the flotation equipment. Surface chemistry is by far the dominant factor in flotation (Leja, 1982). Mineral separation is dependent on both reagent chemistry and water chemistry (Fuerstenau et al., 1985). Flotation equipment (cells) provides the mechanism for air (as bubbles) to come into contact with mineral surfaces so chemical attachment can take place for separation of the selected mineral species. Two types of flotation cells are used in industry today: (1) mechanical flotation cells and (2) column flotation cells. Mechanical cells are by far the predominant type, and except for increasing the size of the process units this fundamental design has not changed significantly for several decades. The design parameters of mechanical cells are fairly well understood, but much is still not known about the design and operation of column cells (Parekh and Miller, 1999).

Mineral separation in flotation requires surface modification for attachment between the mineral and the air bubbles (Ives, 1984). This system requires a careful balance between activators and depressants. Selective flotation is often effected using modifiers to separate gangue minerals from useful minerals. Separation of various sulfides is usually carried out by adjusting the pH of the solution and adding activators and depressants. Pyrite in sulfide deposits is sometimes depressed using cyanide. The introduction of inexpensive, effective, environmentally benign chemical agents would undoubtedly improve mineral separations.

Air is normally used for flotation, but recently nitrogen has been successfully used in flotation for chalcopyrite, molybdenite, and gold (Simmons et al., 1999). Further research will be necessary to determine the potential of this innovation. Unfortunately, the advancement of flotation reagents has been slow. In the past, U.S. chemical manufacturers supported research on less costly, more effective reagents, but in the past 50 years very few flotation reagents have been introduced.

Most flotation is conducted in a water pulp, and yet water, a major component of the system, is probably the least understood aspect of the process. Little attention has been paid to the water used in tests, in spite of the fact that water quality and the ions contained in the water can alter the surface characteristics of minerals, thus having an effect on the separation process (Somasundarum and Moudgil, 1987). The key to the effective separation of fine and ultra-fine minerals may be related to water quality.

During the past decade significant efforts have been made to develop and improve flotation equipment. Large-scale systems and the utility of column flotation cells have been established. Manufacturers' improvements in flotation equipment have been focused on larger units (on the order of 140 cubic meters), but markets for large flotation cells may be limited because of short-circuiting problems and because many operations are small (less that 1,000 tons per day), and the volume of the flotation cell may not be the dominant factor.

In the past 10 to 15 years column cells have been introduced into various flotation circuits, but the use, understanding, and acceptance of these units remain limited. The full potential and understanding of column cells have not been determined. In general, laboratory testing on column cells has not been a reliable method of predicting full-scale plant operation. The scale-up of column cells from laboratory and pilot-plant studies has been problematic and has resulted in plant failures, indicating that not enough is known about how the flotation process interacts with the column-cell dynamics. More developmental research will be necessary on the use of column-cells in operating plants.

New instrumentation has advanced the understanding of flotation fundamentals. Recent research using three-dimensional analysis to examine mineral liberation directly has shown promising results (Lin and Miller, 1997). Research on state-of-the-art instrumentation will require more support to ensure its development and application. Other advances related to flotation chemistry include improved surface spectroscopy, electrochemistry instrumentation, Fourier infrared spectroscopy, and atomic-force microscopy, all of which have improved our understanding of surface-reaction phenomena. Consequently, some improvements in flotation systems have been made, but research is still needed to develop more efficient cell designs and new economical reagents.

Most applications of flotation of industrial minerals are unique in that a large quantity of the incoming material feed to the plant reports to the froth. The separation process is often complex because the minerals in the ore are very similar in composition and crystal structure, such as halite (NaCl) and sylvite (KCl) (Tippin et al., 1999). Therefore, the process requires maximum use of flotation reagents and modifying agents to separate minerals that are similar (Crozier, 1992). Many of the flotation plants in the industrial-minerals industry have multiple circuits with both cleaner and scavenger cells to maximize recovery and produce high-purity concentrates that meet strict market specifications. Beneficial new technologies for the industrial-minerals industry would be: (1) control mechanisms for process parameters; (2) on-stream analysis of mineralogy (not chemical composition); (3) on-stream particle-size or particle-distribution analysis; (4) automation to ensure constant concentrations; and (5) the integration of grinding, classification, conditioning, and flotation unit operations into an understandable model that could be used to operate the process as a single system.

Selective Flocculation

Selective flocculation technology used for industrial minerals is based on the surface chemistry of minerals. In this process chemicals are added to a fine-particle mineral mix resulting in one mineral being flocculated and the remaining minerals being dispersed in a water slurry. Flocculation technologies are used in the iron-ore industry to flocculate and recover iron oxide and in the clay industry to flocculate the quartz and reject grit.

Dewatering

One of the most important aspects of physical separation is dewatering. Once a physical separation is effected in water, the solids and liquids must be separated so unwanted solids can be disposed (Svarovsky, 1977). Thickeners, filters (Orr, 1977, 1979), or centrifuges can be used for dewatering, but final waste products are usually sent to tailings ponds after thickening only. The most desirable final solid products settle quickly and contain minimum amounts of water. Various chemicals are used to achieve this.

Hydrometallurgy and Chemical Processing

Hydrometallurgy encompasses leaching in water with various chemicals, assisted by oxidizing agents, elevated oxygen partial pressure, or bio-oxidation; dissolved species are removed by precipitation, solvent extraction, electrowinning, or adsorption. Leaching may be carried out in vessels, heaps, or dumps. Unlike physical separations, hydrometallurgy is capable of yielding solutions of relatively pure metal ions, which usually can be recovered directly. Hydrometallurgy has become increasingly important over the years and is now a major aspect of extractive metallurgy.

Heap Leaching and Dump Leaching

The development of heap-leaching and dump-leaching technologies for low-grade ore has extended the world's ore resource base considerably. The processes were developed by the copper industry and the USBM and have been extended to uranium and gold. The gold industry adopted innovations, such as feed preparation (agglomeration), heap design and construction, and solution distribution. Bioleaching for heaps and dumps was developed by the copper mining industry and adapted to treat refractory gold. To speed up leaching rates and improve recovery rates, successful heap-leaching and dump-leaching operations require a combination of geology, mineralogy, hydrometallurgy, hydrology, and modeling, and sometimes biology (when bioleaching is used). Solution management and understanding the characteristics of the ore to be leached are key elements of heap-leaching and dump-leaching operations.

Mathematical modeling to profile metals and predict optimum performance would improve the overall rate of metals recovery. Recently, encouraging results have been obtained using a high-resolution resistivity technique to survey poorly wetted (nonpenetrated) areas in the heap. Selective releaching of the resistive zones has increased the recovery rate. In-situ mining aids, such as catalysts, surfactants, and wetting agents, may accelerate leach kinetics and increase the permeability of rock surfaces. An important advance in heap leaching and dump leaching would be the development of new lixiviants that could effectively extract metals directly in an environmentally friendly manner, especially metals from refractory sulfide minerals (Sparrow and Woodcock, 1995). New lixiviants would be particularly beneficial for maximizing metals extraction from near-surface deposits using in-situ techniques.

High-Pressure Technology

The use of high-pressure technology has been demonstrated for a variety of commodities in acidic and basic solutions under oxidizing and reducing conditions. Because of increased reaction rates for both oxidative and reductive processes, pressure hydrometallurgy would be a suitable technology for the future. Large-scale autoclaves are used for production of zinc and nickel and to treat refractory gold ores (Mason and Gulyas, 1999). New developments in autoclave technology for pressure leaching copper concentrates may also be useful for other mineral systems.

Recently developed processes include pressure leaching with ultra-fine grinding, which increases leaching kinetics and metal recovery, thus making pressure hydrometallurgy more attractive. Selective leaching and precipitation capabilities are available at elevated temperatures, which may be achieved at elevated pressures. Improved gas/solid/liquid mixing in the autoclave and development of catalysts and other chemical aids are becoming increasingly important to accelerating reaction kinetics.

Bioprocessing

Bioprocessing, the application of biotechnology to the extraction and recovery of metals, is becoming an increasingly important hydrometallurgical processing tool. Bioprocessing is divided into (1) bioleaching/mineral bio-oxidation technology and (2) biotechnology for the recovery and concentration of metals from aqueous solutions. Bioleaching uses the catalytic properties of micro-organisms to dissolve metals into an aqueous solution. An example of bioleaching is the microbially catalyzed oxidation of chalcocite to solubilize copper in acidic water. Mineral bio-oxidation is a pretreatment process that uses micro-organisms to catalyze the oxidation of a sulfide mineral, such as pyrite, exposing precious metals for subsequent dissolution by another reagent, such as cyanide. Another aspect of bioprocessing involves the removal of metals from a solution, using micro-organisms themselves or products of micro-organisms to concentrate or immobilize them.

Bioleaching and mineral bio-oxidation are in commercial use today (1) in dumps to scavenge copper from run-of-mine rock; (2) in heaps to leach copper from secondary copper ores and to pretreat precious-metal ores in which the gold and silver are locked in a sulfide-mineral matrix; and (3) in aerated, stirred-tank reactors to pretreat precious-metal concentrates and bioleach base-metal concentrates (Australian Mineral Foundation, 1999; Amils and Ballester, 1999a, b; Rawlings, 1997). Bioleaching/mineral bio-oxidation employs a variety of different microorganisms that function under acidic conditions and a range of temperatures. Despite its commercial uses little is known about the microbial ecology of the heaps, dumps, aerated reactors, or the microbial/mineral interactions that occur in these systems.

The principal drivers of the development and commercial application of bioprocessing are (1) decreased production costs through lower energy requirements, lower reagent usage, and lower labor requirements; (2) lower capital costs because of simpler equipment and faster construction; (3) an increased reserve base because lower grade ores can be economically processed and more diverse mineral types can be processed; and (4) improved environmental conditions and worker safety because there are no gaseous emissions and, in some cases, no aqueous discharges. Research on fundamentals and applications of all aspects of bioprocessing would significantly improve the effectiveness of the technology in current applications and extend its application to in-situ leaching and metals recovery and concentration.

Genetic modification using standard "adaptation and selection" techniques is usually used for micro-organisms employed in aerated, stirred-tank reactors to ensure that microbial cultures can tolerate high metal concentrations. Modern molecular-biology techniques have not been used to genetically modify microorganisms used in commercial practice but have been used to identify and enumerate micro-organisms in heaps and stirred-tank reactors. The use of modern molecular biology to modify microorganisms genetically for processing applications should be pursued as "blue-sky" research. Because mineral processing applications release organisms into the environment, researchers will have to be mindful of regulations governing the release of genetically engineered organisms into the environment.

The solubilization of metals from mineral matrices can also be accomplished with a large number of micro-organisms other than the acid-loving ones currently used commercially. Bioleaching under neutral or slightly alkaline conditions has been carried out with bacteria, fungi, yeast, and algae through several metabolic means. Most studies on nonacidic bioleaching have focused on understanding microbial processes as they relate to subsurface microbiology, rock weathering, and mineral deposition, as apposed to hydrometallurgical processing. Research on the most promising nonacidophilic microbial processes for metals mobilization coupled with careful process-engineering design and rigorous economic assessments of the proposed processes could result in commercially usable technologies that are more energy efficient and less polluting than the technologies used today.

Bacteria, fungi, yeast, and algae recover and concentrate metals from solutions using a variety of metabolic strategies. Most research in this area is focused on using microbial mechanisms for environmental management as opposed to hydrometallurgical processing. Considerably more research and development will be necessary before this area of bioprocessing delivers commercially viable products and processes. A more fruitful approach might be to focus on developing these technologies for environmental control in mining and, when demonstrated effective for that use, modifying the technology for metallurgical processes.

Solution Purification and Concentration

The vitally important hydrometallurgical processes of solution purification and concentration not only concentrate metal ions from dilute leaching solutions to levels suitable for metals recovery but also selectively reject other impurities. Over many years chemists learned how to precipitate dissolved metals from solution by a number of techniques, and by the early 1800s they suspected that some sort of electrical phenomenon might be involved. Miners knew well before that time that strong acids could dissolve some of the metals in an ore and that copper could be collected in solid form by "cementation." This technique survived until the 1950s, although "cement" copper was not pure enough for electrical uses. Since then the cementation process has been superseded by solvent extraction and electrowinning (SXEW).

Technologies used for solution purification and concentration include precipitation, ion exchange, solvent extraction, membrane transfer, and electrowinning. An excellent discussion of these methods and others can be found in *Separation Technologies for the Industries of the Future* (NRC, 1999a). All of them have been used for nonmining chemical separations, and precipitation has been used in the leach-precipitation-flotation process for copper.

Solvent extraction is a phase-transfer process between organic and aqueous phases. The process, which was well known to analytical chemists, became an industrial process during World War II, when it was used for such separations as zirconium/hafnium, uranium/vanadium, and plutonium purification. The first commercial solvent-extraction process for uranium was installed in 1955 at the Kerr-McGee plant in New Mexico. In 1963 General Mills succeeded in producing a copper reagent, LIX 63, which in combination with electrowinning led to the first small copper SXEW plant in Arizona. The development of solvent extraction made copper hydrometallurgy the successful process it is today. The use of solvent extraction has been extended considerably to the production of nickel, cobalt, and rare-earth elements.

Stable emulsions and the eventual formation of "crud" are problems common to most solvent-extraction operations in the mining industry. Crud can constitute a major solvent, uranium, and copper loss to a circuit and therefore adversely affect the operating cost. Overcoming solvent loss and improving the rate of metal recovery will depend on the development of new extractants, modifiers, and diluents. Solvent-extraction methods could be extended to other applications with the development of a larger suite of selective reagents. The design and operation of mixer-settlers for optimization of solvent-extraction performance and entrainment minimization could also be improved.

Ion exchange is a mature technology used in many industries. The application of ion-exchange technology to hydrometallurgy began with uranium extraction. The technology is based on resin beads containing exchangeable ions or groups and is employed in columns (for clear solutions) or directly in the pulp (resin-in-pulp). Ion exchange is uniquely suited to extraction from very low-grade (ppm) solutions where losses using solvent extraction would be excessive. Ion exchange also eliminates transport of flammable diluents. Challenges and great opportunities occur in the development of selective resins strong enough to withstand rough handling. Currently, considerable attention is being focused on increasing selectivity by introducing chelating functional groups into the resin.

Membrane technology and extractant-impregnated membranes/porous particulates are also important phase-transfer processes for the future. Thin-film, polymeric membranes with separation characteristics between reverse osmosis and ultra-filtration became available in the mid-1980s. In the past five years membrane-separation technology for concentrating ions from very dilute solutions has been developed for wastewater-treatment applications. Certain types of multilayer, thin-layer, polymer composite membranes that are stable in strongly acidic environments are capable of making separations among monovalent, divalent, and trivalent ions. However, like most technology developments, there are gaps between research, technology transfer, and implementation of the new process.

Solid-liquid separation is an important part of nearly all hydrometallurgical operations. Large amounts of gangue solids must be separated from the pregnant solution after leaching, and a clarified solution is required for downstream metal recovery. The economics of a hydrometallurgical plant are often influenced by the cost of solid-liquid separation.

Iron precipitation has played an important role in many chemical reactions. The formation of jarosite-type compounds, which are complex alkali iron sulfates, provides a ready avenue for the precipitation and elimination of alkalis, iron, sulfate, and other impurities from chemical processing solutions (Das et al., 1996). One major advantage of precipitating jarosite-type compounds is the comparative ease of settling, filtration, and washing of the resulting solids. Further research with iron precipitates will be necessary to minimize the amount of precipitate formed, to enhance coprecipitation of other deleterious impurities with the iron to improve downstream processing for the metals of value, and to develop better methods for final disposal of the iron precipitates.

One of the oldest and cleanest metal-separation technologies used in hydrometallurgy is electrowinning, an energy-intensive technology. Major advances in electrowinning have been made, but further improvements can be expected if overpotentials can be reduced, current densities increased, and at the same time morphology and purity of depositing phases controlled. Improvements in anodic and cathodic process controls, as well as minimization of acidic or toxic mist in the tank house, are important goals for the future.

Hydrometallurgical process streams are by nature very complex. The complexity and harsh environments characteristic of process streams make online instrumentation difficult. Sophisticated online sensors and instrumentation should be based on the chemistry and operating parameters of the individual operating unit. The main requirements of sensors are robustness, reliability, and operability in the process environment. Sophisticated online sensors for critical variables of some hydrometallurgical processes could be optimized. The research challenges are the complexities of structuring a well-coordinated control and optimization scheme. Although control mechanisms, mostly implemented with computers, have improved rapidly in recent years, the industry could benefit from accelerated research on sensors. Progress on sensors and controls could increase the efficiency and productivity of existing processing facilities.

All mineral processing technologies have an effect on the environment. When valuable metals are dissolved in aqueous media, other metals may dissolve as well. Some of these metals are regarded as toxic and hazardous and must be removed before effluents can be discharged to the environment. Therefore, the development of innovative, environmentally friendly technologies will be extremely important. Minimizing waste generation and using wastes to produce useful by-products while maintaining economic viability must be a goal for new technologies. Hydrometallurgy uses the most sophisticated aspects of kinetics, solution chemistry, and electrochemistry to realize its full value. Major advances in understanding fundamental chemistry and physical phenomena in processing will contribute to improved extraction and separation efficiencies, as well as minimize environmental impact.

Hydrometallurgy technology is not used extensively in the processing of industrial minerals, almost all of which are insoluble in normal, low-cost acidic and basic lixiviants. Leaching or other hydrometallurgical techniques are used only for industrial minerals to remove minor contaminants of a product or to prepare the mineral surface to meet market specifications.

A few industrial minerals are "surface treated" for special industry applications. To meet market specifications in the kaolin and clay industries products are often bleached with a sodium hydrosulfite or similar compound to improve whiteness or brightness, or ozone is added to oxidize organic substances. Other minerals, such as mica, are surface treated with organic compounds to achieve selected coatings on the mineral. Hydrometallurgical techniques are also used in the production of lithium, boron, soda ash, sulfur, and other unique minerals. However, these minerals require specialized chemical processes. In general, the industrial-minerals industry has very limited interest in hydrometallurgical research; its research needs are either specific to a single mineral or are required to minimize adverse environmental impacts. The application of hydrometallurgical techniques in the coal industry is even more limited. The coal industry may, however, be interested in the chemical removal of mercury or sulfur, or both.

Recommendations for Processing Technologies

Research and development would benefit mineral processing in the metal, coal, and industrial-mineral sectors in many ways. Every unit process—comminution, physical separation, and hydrometallurgy/chemical processing—could be improved by technical input ranging from a better understanding of fundamental principles to the development of new devices and the integration of entire systems.

Because comminution is so energy intensive, the industry

would significantly benefit from technologies that improve the efficiency of comminution (e.g., new blasting and ore-handling schemes) and selectively liberate and size minerals. Fine-particle technologies, from improving production methods for the ultra-fine grinding of metals to minimizing the production of fine particles in coal preparation and measuring and controlling the properties of industrial mineral fine particles would be useful.

Technology needs in physical-separation processes are focused mainly on minimizing entrained water in disposable solids, devising improved magnetic and electrostatic separators, developing better ore-sorting methods, and investigating selective flocculation applications. Although flotation is a well-developed technology, the mining industry would benefit from the availability of more versatile and economic flotation reagents, on-stream analyses, and new cell configurations.

The most important change in the mineral industry in the next 20 years could be the complete replacement of smelting by the hydrometallurgical processing of base metals. This development could be the continuation of a trend that began with dump leaching and heap leaching, solvent extraction/electrowinning, followed by bioleaching and pressure oxidation. Future research and development focused on innovative reactor designs and materials, sensors, modeling and simulation, high-pressure and biological basics, leaching, and separation reagents are likely to continue this trend. Like other components of mining, mineral processing could also benefit from the integration of unit processes for optimal performance, economic benefits, and environmental benefits. Research and development opportunities specific to mineral processing are listed in Table 3-4.

TABLE 3-4 Opportunities for Research and Development in Mineral Processing

COMMINUTION

Metal and Coal Mining
- effects of various blasting and ore-handling schemes on comminution efficiency and productivity
- more energy-efficient comminution methods
- more energy-efficient production of ultra-fine particles
- improved components leading toward overall optimization of mining systems
- development of fast, accurate, mineral-liberation analysis with feedback to comminution
- minimization of fine particles in coal production
- development of more efficient recovery and reconfiguration of coal fines
- improved wear resistance in materials used in crushing and grinding

Industrial Minerals
- improved methods for selective grinding and efficient sizing
- improved quality control for consistently sized products
- measurement and control of product size
- measurement of properties of dry fine particles

PHYSICAL SEPARATION

Gravity Separation
- efficient gravity separation for fine particles and ultra-fine minerals
- improvements in multiforce separation technology
- more efficient heavy media
- integration of approaches in multiprocess (gravity and flotation) systems for total plant optimization
- lower cost and increased dependability of process instrumentation and control mechanisms for process parameters

Magnetic/Electrostatic Separation (Wet and Dry)
- efficient, economical, high-intensity magnetic separation technology
- expansion of efficient, economical electrical separation technology
- broader use of efficient, economical nonconductive (eddy-current) type separation technology
- improved on-stream mineral analysis and particle-size distribution

Ore-Sorting Separation
- discovery of specific reagents or instrumentation for surface identification of individual minerals
- ore-sorting systems for minerals with particle sizes under 2.5 mm
- methods of rapid, continuous mineral identification

Flotation
- improved flotation systems for fine and ultra-fine mineral systems
- improved, economical flotation reagents
- improved understanding of column-cell process dynamics and cell design
- perfected on-stream analysis of mineral species, chemical composition, and particle-size distribution
- optimized flotation operations

Selective Flocculation
- selective flocculation processes

HYDROMETALLURGY AND CHEMICAL PROCESSING
- nontoxic, efficient lixiviants for metals extraction
- control and management of environmental hazards and stabilization of solid wastes and aqueous effluents
- new corrosion/abrasion-resistant materials for chemical-processing reactors
- robust, effective on-stream sensors
- models and simulations of processing to predict and optimize processing
- fundamentals of high-pressure and high-temperature reactions
- selective, stable ion-exchange resins and polymers for metals separation
- membrane technologies

Biotechnology
- fundamental advances in understanding of micro-organism/mineral interactions, genetics related to monitoring microbial activity in processing and strain development, and nonacidic microbial leaching systems
- basic research on the genetics of micro-organisms used in mineral processing to develop enumeration and identification techniques and improve microbial strains
- identification of nonacidic microbial mineral-processing technologies with scale-up of the most technically and economically promising processes
- bioprocessing methods for selective metal recovery and concentration

4

Health and Safety Risks and Benefits

Chapter 3 identified technologies that will benefit major components of the industry in the areas of exploration, mining, and processing. This chapter discusses concerns for potential new health hazards arising from the introduction of new technology.

At the beginning of the twentieth century approximately 3,000 coal miners and 1,000 metal/nonmetal workers were killed annually in mine accidents (Ramani and Mutmansky, 2000). In 1910 the U.S. Congress created the USBM in the U.S. Department of the Interior to conduct research on the health and safety problems associated with the extraction and processing of coal and minerals. Since that time, overall safety conditions in mines have improved as a result of several factors, including the pioneering research by the USBM on hazards identification and control, major improvements in mine design, the passage of stringent health and safety legislation, and the introduction of more productive systems. Cooperative efforts on the part of industry and the USBM led to advances in mining sciences and technology that dramatically reduced deaths and injury-causing accidents (Figures 4-1 and 4-2) (Ramani and Mutmansky, 2000).

The declining trends in fatal and nonfatal incidents have continued to this day (Figures 4-3 and 4-4). Major changes that have influenced these trends are the prevalence of surface mining, larger mining equipment and record-setting mine production and productivity rates (Adamczyk, 2000; Phelps, 2000; Ramani and Mutmansky, 2000; Carter, 1999; Shuey, 1999; Wheeler and Walls, 1998; Hartman, 1987).

In the last three decades industry-specific health and safety legislation has progressed from the state level to the federal level. The 1969 Coal Mine Health and Safety Act and the 1977 Mine Safety and Health Act were landmark laws that brought significant advances in mine health and safety. As a consequence, at the end of the twentieth century yearly coal mine fatalities had dropped to one-sixtieth and metal/nonmetal mine fatalities to one-twentieth of the rates at the beginning of the century; at the same time, production today dwarfs production then.

According to the National Safety Council statistics of selected industries, safety in the mining industry is near the top one-third of industries reported (National Safety Council, 1999). Major disasters have been effectively controlled, although death and disabling injuries continue to occur from machinery, roof falls, and electrical accidents. Nevertheless, the fact that fatalities and injuries continue to occur is a cause for concern.

On the health front, miners have long been aware of the hazards posed by the gases, dusts, chemicals, and noise in the work environment and in working in extreme temperatures (hot or cold) and at high altitudes. Silicosis, pneumoconiosis (black lung disease), occupational hearing loss, and other medical problems have long been associated with mining operations. The 1969 Coal Act established, for the first time, stringent requirements for the control of airborne, respirable coal-mine dust as a means of reducing incidences of pneumoconiosis and silicosis, and major reductions have been made in airborne, respirable dust concentrations in underground coal mines (Figure 4-5).

Although health conditions in coal mines have improved considerably since 1969, new cases (almost 150 in 1999) of these lung diseases are still being reported (U.S. Department of Labor, 2000b). At the present time, about 40,000 former miners are disabled from lung diseases, such as black lung disease and silicosis (U.S. Department of Labor, 2000b). Potential health hazards introduced by new technologies must be addressed proactively because they may not be immediately apparent. For example, the use of diesel equipment in the mine environment has raised concerns because of the presence of other contaminants in the mine atmosphere.

Chemical and biological hazards must also be given more attention. The use of chemicals in many areas of mining, particularly in hydrometallurgy and in-situ leaching, is also increasing. The introduction of new lixiviants in hydrometallurgy and bioagents to facilitate the leaching of metals may create new health and safety hazards.

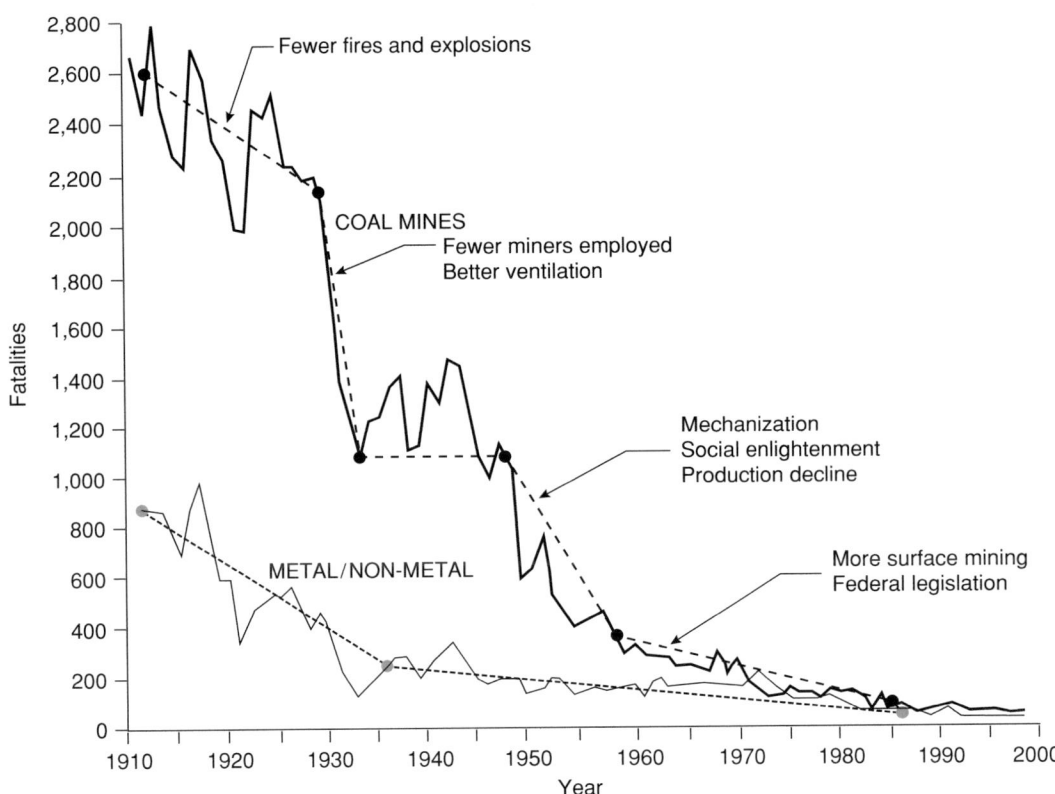

FIGURE 4-1 U.S. mine fatalities, 1910 to 1999. SOURCE: Katen, 1992; Ramani and Mutmansky, 2000; U.S. Department of Labor, 1999.

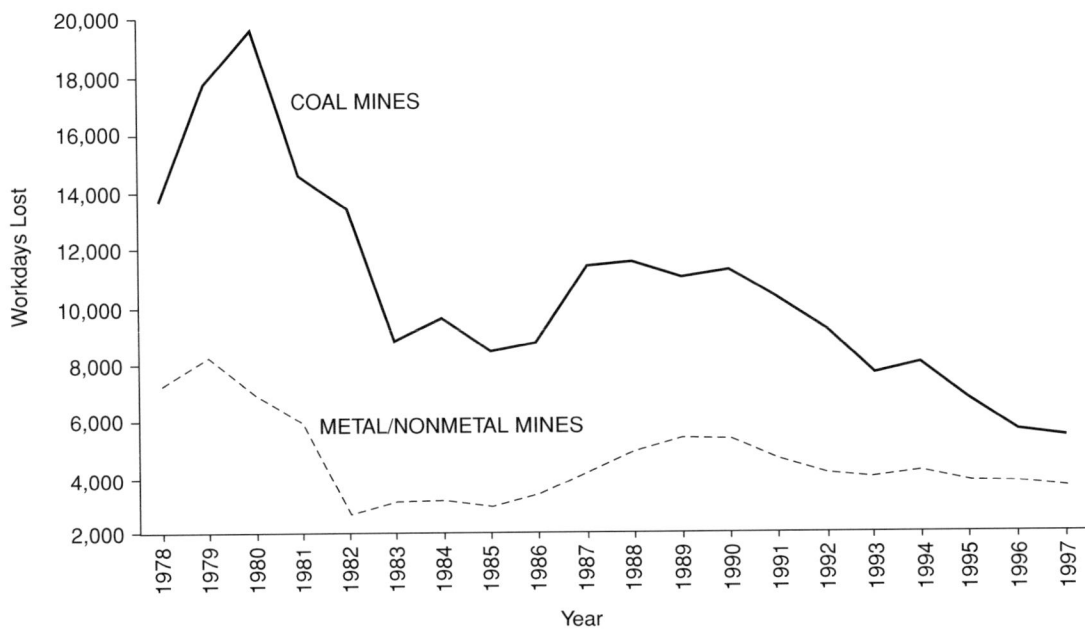

FIGURE 4-2 Nonfatal lost workdays, 1978 to 1997. SOURCE: Raja Ramani and Mark Radomsky.

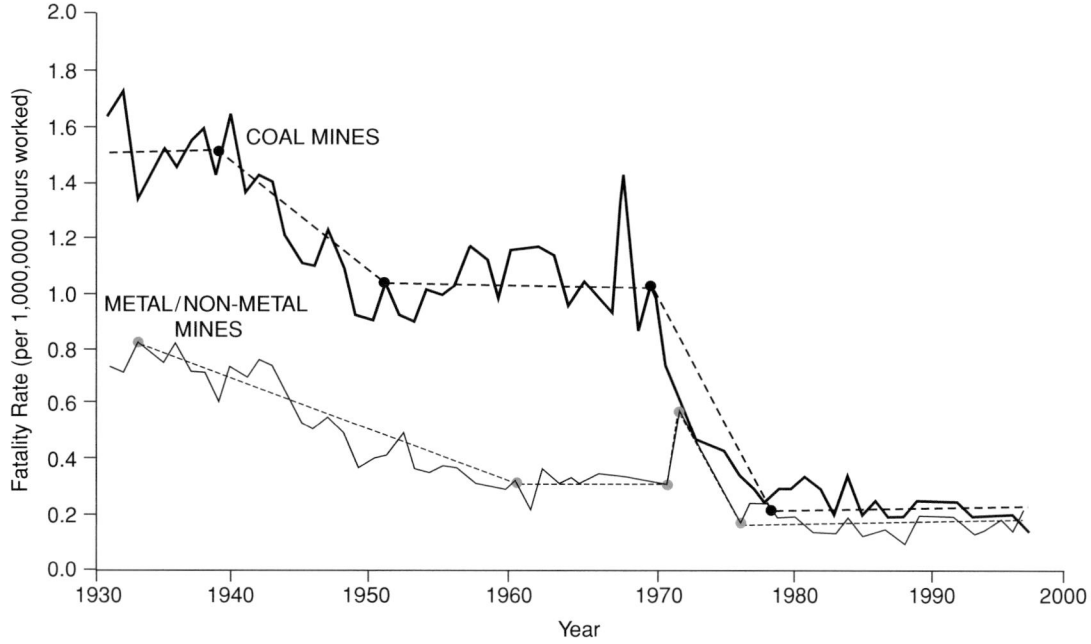

FIGURE 4-3 U.S. fatality rates, 1931 to 1999. SOURCE: Katen, 1992; Ramani and Mutmansky, 2000; U.S. Department of Labor, 1999.

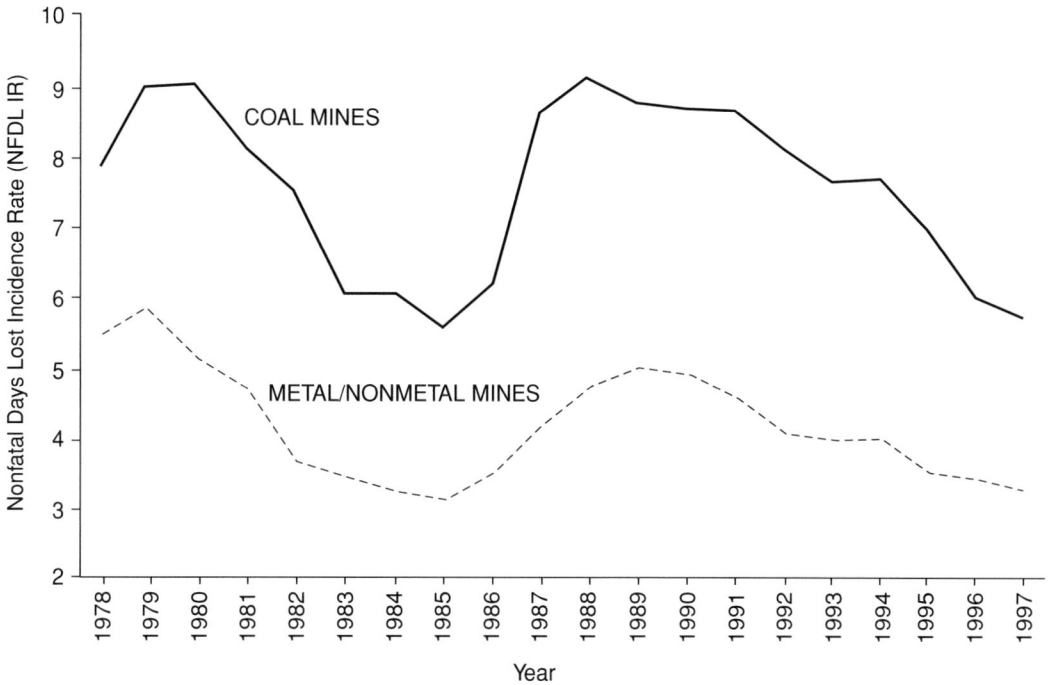

FIGURE 4-4 Nonfatal days-lost rates, 1978 to 1999. SOURCE: Raja Ramani and Mark Radomsky.

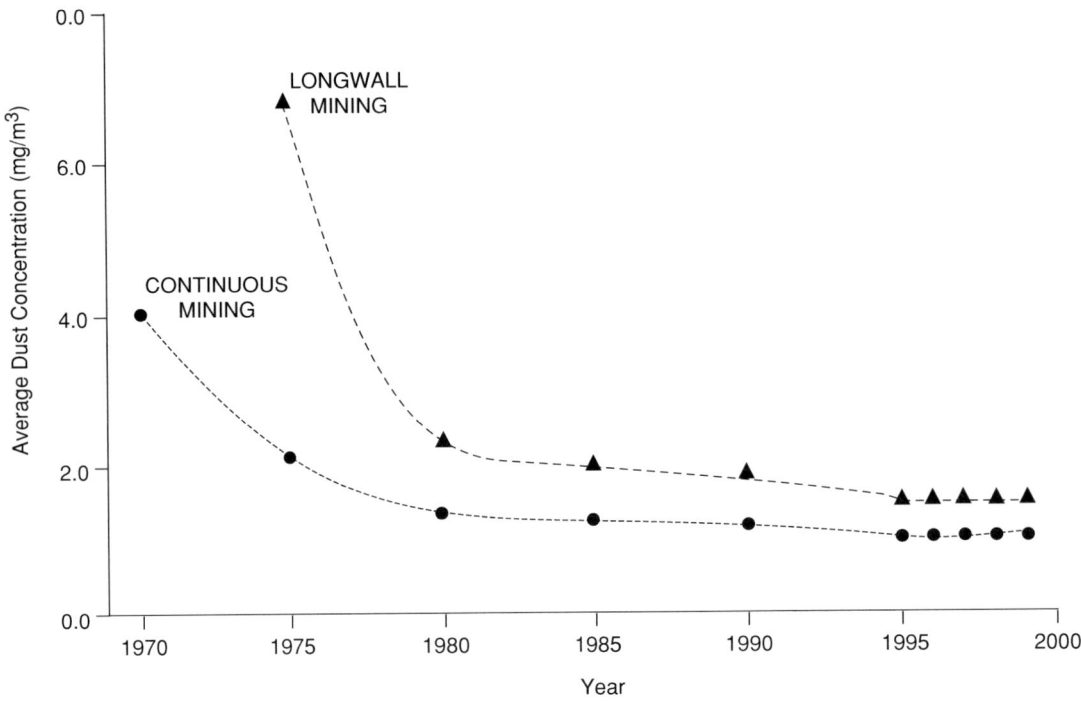

FIGURE 4-5 Average dust concentrations for U.S. longwall and continuous mining operations. SOURCE: Gillette et al., 1988; Ramani and Mutmansky, 2000.

SIZE OF EQUIPMENT

One technological trend in the mining industry is the steady increase in the size of mining equipment. As the level of automation increases, large machines may be able to operate in a fully automated mode (Shuey, 1999). Larger equipment will certainly decrease individual exposures to hazardous conditions in the mine environment. However, it will also introduce new problems. For example, the increasing size of mobile mining machinery, particularly large hauling equipment, has increased the likelihood of accidents by decreasing all-around visibility for the operator. If the operator does not know his precise location in the mine, near a berm or where nearby objects, such as other miners and smaller equipment are, fatal and costly accidents can result. Technology that can alert equipment operators to the presence of obstructions (e.g., other equipment, berms, miners) and their relative distances would mitigate this problem. In addition to the extensive use of closed-circuit television, automatic control of the equipment, aided by onboard sensors and GPS monitors, could reduce these hazards.

AUTOMATION

Several mining systems (e.g., longwall systems in underground coal mining) are already highly automated. Semiautonomous and fully autonomous systems can result in higher levels of production and productivity, as well as better health and safety conditions. However, automated equipment, even remotely controlled equipment, can create new hazards. Because the number of automated systems in the mining industry is small, no extensive data on health and safety are available. However, it is known that automation has been responsible for a small number of mine accidents involving deaths and disabling injuries (NIOSH, 1999).

Automated equipment is also subject to unexpected or unplanned movements. Unique challenges presented by the mining environment to the designer and operator of automated equipment include the lack of precise knowledge of operating conditions, mostly because of variations in geology; a demanding operating environment exacerbated by extremely dusty, noisy, and vibrational conditions; specialized equipment design to meet the stringent safety standards required to operate in mines; and more rigorous maintenance programs for proper operation. Research could focus on understanding the design specifications for automated operation in the mining environment and ensuring the robustness and reliability of equipment and systems. At the same time, new training programs will be necessary to address new hazards. Virtual reality systems for training miners and operators would be useful.

ERGONOMICS

Back injuries, especially strains, are common among miners (NIOSH, 2000). As thinner seams and narrower veins are mined, new technologies and more ergonomic designs will be necessary to reduce the number of injuries and to enable workers to operate in confined, awkward spaces. Workplace risk factors that lead to musculoskeletal disorders (e.g., low back pain) should be identified and taken into account in the design of equipment and training programs. Remote control and autonomous operation would eliminate the problem.

ALTERNATIVE POWER SOURCES

Alternative power sources are being explored for increasing the efficiency of mining systems and reducing hazards from electrical power. Although hydrogen-powered vehicles are under development, the use of diesel-powered equipment has been on the rise. The benefits of research on cleaner, more efficient diesel engines for other applications (e.g., surface highway vehicles) will be useful to mining equipment manufacturers.

Safe operation of diesel-powered equipment in mines is affected by the gas, dust, visibility, vibrations, noise, and degree of confinement. The impact on the health and safety of miners requires research on a number of fronts: the development of monitoring equipment that can discriminate among the sources of airborne pollutants (blasting, diesels, oxidation, cutting, etc.) and an instrument that can reliably measure the amount of particulates in the mine. A related need is for accurate, real-time monitors of personal exposure. For example, a device that indicates in real time the level of a miner's exposure to respirable coal-mine dust (personal exposure) will be essential to controlling excessive exposures. Another area for study is health effects of mixed-mode exposures to mineralogical constituents of the mined materials and particulates in diesel exhaust.

NOISE

Audiometric data on coal and metal/nonmetal miners shows that noise-induced hearing loss in miners has become a substantial problem (U.S. Department of Labor, 1999). The Mine Safety and Health Administration (MSHA) recently issued the final rule for occupational noise exposure. Risk of hearing loss for an individual increases with the level of noise and the duration of the exposure. Minimizing the risk of hearing loss must be an essential aspect of new technology development. Research should focus on the development of materials that reduce noise levels (source control) and remote, automatic control technology for noisy mining equipment. Training miners in safe operating procedures in a noisy environment and the correct use of personal noise protection devices will also be necessary.

COMMUNICATIONS

Research linking computer-oriented monitoring of conditions in mines with a safety information system and a rapid communication system for transmitting that information in real time to miners could have tremendous payoffs for the entire industry. As equipment and mining systems become increasingly automated, miners may not be working in close proximity or in visual contact with each other. In fact, most miners may be in remote areas of the mine. Therefore, the need to communicate with each miner, based on real-time data and analyses, will be critical. Some components, such as atmosphere-monitoring systems, equipment-monitoring systems, and pager systems, have already been developed. These advancements must now be integrated so specific, immediate safety information and instructions can be communicated to individual miners.

TRAINING TECHNOLOGY

Any change in the mining system — a new environment, new equipment, new rules, or new personnel—requires that miners be trained to accommodate it. Mandatory health and safety training would be greatly improved by computer-oriented training tools and techniques for miners and supervisors. Computer-based training would use the capabilities of computers to overcome some of the limitations of traditional classroom and on-the-job training programs. Computer-based training would also promote better designed and higher quality lessons and self-learning. Virtual-reality training modules would improve miners' ability to react appropriately to hazardous situations.

RECOMMENDATIONS

Advances in technology have historically improved the health and safety of miners. Relatively new technologies, such as in-situ mining and automation, can significantly reduce exposures to traditional mine hazards. Increasing production and productivity with larger equipment can also

TABLE 4-1 Recommendations for Research and Development in Health and Safety

- technology to alert equipment operators of the existence and location of obstructions (such as equipment, berms, miners)
- design specifications for automated operation in the mining environment that enhance robustness and reliability
- miner-training programs to address special hazards that are created by the introduction of automated systems
- identification and elimination of workplace hazards introduced by new chemicals and bioagents
- identification of workplace risk factors that lead to musculoskeletal disorders (e.g., low back pain) and the design of equipment and training programs to eliminate them
- technology for assessing health and safety conditions in mine atmospheres; monitoring equipment that can distinguish the sources of airborne pollutants (blasting, diesels, oxidation, cutting); an instrument that can reliably measure the amount of diesel particulate matter; instruments that can accurately measure real-time personal exposures, particularly exposures to airborne respirable coal-mine dust
- determination of the health effects of mixed mode exposures in mine environments
- new materials and technologies to reduce noise in mining equipment and systems
- linking of computer-oriented monitoring of conditions in mines with a safety information system and a rapid communication system to provide specific information in real-time to each miner
- virtual-reality training modules for miners and mining-equipment operators

reduce exposures to health and safety threats. At the same time, these advancements will certainly introduce new hazards and, in some cases, may exacerbate known hazards.

New monitoring and control systems could effectively address the mining-equipment and mine-system safety issues. Advances in industrial training technologies, for example, have immense potential for improving safety. Most of these advancements would be realized through intelligent combinations of sensors, analyses, visualization, and communication tools that will either eliminate a hazard or enable a miner to take rapid actions to avoid an emerging hazard. Several areas for research and development are identified in Table 4-1.

5

Research Opportunities in Environmental Technologies

INTRODUCTION

Chapter 3 identified technologies that would benefit major components of the mining industry in the areas of exploration, mining, and processing. Environmental risks and benefits of some of the technologies discussed in Chapter 3 were also addressed there. This chapter describes research opportunities and technologies that would lessen the environmental impacts of mining.

The creation of large-scale surface disturbances, the generation of large volumes of waste materials, and the exposure of previously buried geologic materials to the forces of oxidation and precipitation are intrinsic to the mining industry and may continue to present complex environmental problems even when the best available practices are conscientiously followed (Chiaro and Joklik, 1998). Even with all of the recent advances in environmental protection, the mining industry still cannot completely and effectively predict, prevent, or treat releases from mine sites during and after operations.

Technologies that predict, prevent, mitigate, or treat environmental problems will be increasingly important to the economic viability of the mining industry. With new and effective environmental technologies new mining operations may be permitted that might otherwise have been rejected because of unacceptable environmental risks. New technologies may also result in reduced costs for environmental compliance and facility closures. Improved environmental technologies related to mine closures have the greatest potential for increasing overall productivity and reducing overall energy consumption because costs of long-term maintenance will be factored into the analysis. Closure procedures usually affect the quality of drainage water from the mine and thus, productive use of the land. Long-term (*in perpetuity*) treatment of drainage water from mines now requires capitalization by interest-bearing funds. Today, perpetual treatment or maintenance often appear to be economically preferable to expensive permanent treatments. Future technologies that would allow "walk-away" closures by preventing drainage problems would have obvious environmental and economic benefits. Moreover, permanent solutions prevent problems from becoming exacerbated over time rather than simply managing their impacts. The industry now generally recognizes that planning for closure must begin when the mine is in the planning phase.

RESEARCH OPPORTUNITIES AND TECHNOLOGY AREAS

The mining of coal, base and precious metals, and industrial minerals raises several environmental issues. Some are common to all mining sectors; others are specific to one sector or even to one commodity within a sector (Sidebar 5-1). The differences may be attributable to differences in the geologic settings in which the materials are found, differences in the regulatory regimes under which they are mined, and/or differences in production processes. For example, environmental planning in surface coal mines and surface areas of underground coal mines must meet the performance standards of the Surface Mining Control and Reclamation Act of 1977, which has established standards for closure and reclamation.

Even before the federal law was passed, several coal-mining states, particularly in the eastern United States, had longstanding, mining-specific legislation requiring the reclamation of disturbed lands. As a result of this long history of mandated standards, industry and governmental agencies have made considerable efforts to address the problems specified by the regulations, specifically air quality, water quality, land use, and noise. Industrial-minerals operations in close proximity to communities have been subject to local zoning regulations, which has had a major influence on planning, including environmental planning. Even if the problems facing all mining sectors are similar (e.g., lakes in the final pit, tailing ponds, and waste and spoil piles), the magnitude and severity of the problems and the applicable

> **SIDEBAR 5-1**
> **Phosphogypsum**
>
> Disposal of phosphogypsum is a special problem. The production of phosphate generates 5 tons of phosphogypsum for every ton of phosphate product manufactured. In Florida alone nearly a billion tons of phosphogypsum has been generated. This by-product is generated when phosphoric acid is manufactured using sulfuric acid, an inexpensive reagent. The resulting phosphogypsum must be managed as a waste. Other acids could be used to manufacture phosphoric acid; however, the by-products of these processes are an insoluble calcium salt, which also raises disposal issues. Although potential uses for phosphogypsum have been identified, its use is prohibited by the Environmental Protection Agency because of the presence of radium at a concentration greater than 10 picocuries per gram present in Florida phosphogypsum. Potential uses of phosphogypsum include material for road bases, agricultural fertilizer to supply sulfur and calcium, and a supplement in landfills and sewage treatment plants to enhance the microbial decomposition of organic waste.
>
> Research to identify or develop uses for phosphogypsum that contains radium could reduce the amount of material that must be managed as waste, as well as limit its potential environmental impact. Research should also focus on the development of a phosphate manufacturing technology that does not produce phosphogypsum as a by-product.

solutions may not be similar. Nevertheless, all mining sectors would benefit from new technologies to reduce or eliminate adverse environmental impacts caused by mining operations.

Water-quality issues related to mine closures are often the most challenging and costly to address for all types of mining. They also present significant opportunities for research that could increase overall productivity. New technologies for managing other materials of environmental concern, such as innovative, cost-effective solutions to managing slimes, could also increase productivity in the mining industry.

Acid-Rock Drainage[1]

The most common water-quality problem associated with metal and coal mining is the release of metals, acidity, and sulfate from reactive rock surfaces, particularly when pyrite is present (University of California, 1988). Sulfide-containing rock is oxidized when newly fractured rock comes in contact with oxygen, water, and (very often) bacteria. The sulfides are converted to sulfuric acid, which in turn can mobilize a variety of metals (Alpers and Blowes, 1994; Doyle and Mirza, 1990). The most common metals and related elements released are iron, copper, zinc, cadmium, manganese, arsenic, antimony, lead, nickel, and mercury. The concentrations of metals vary dramatically with the type of deposit and the environmental factors surrounding the site.

The pH of the water can also vary, depending on the amount and type of sulfide minerals oxidized, as well as the amount of neutralization capacity present in the rock. Highly acid-generating rock produces a drainage pH of less than 1; well buffered-rock (e.g., rock that contains calcite or dolomite) may have elevated sulfate concentrations indicative of sulfide oxidation but still retain a near-neutral pH. Metal concentrations, and thus the environmental and health impact of drainage water, depend on the pH. Low pH water (high acidity) has much higher concentrations of metals than near-neutral pH water (low acidity). Some potentially toxic elements, such as selenium and arsenic, are problematic at any pH.

Acid generation is an issue of concern for certain types of coal and hardrock mines (particularly mines that contain high concentrations of pyrite but low concentrations of pH-buffering materials). Problems are most common for reactive waste-rock dumps, although tailings and waste facilities, low-grade ore piles, highwalls, and precious-metal heaps can also generate acids under specialized conditions. Acidic drainage into receiving waters can result in severe impacts to the biological integrity of a stream and can change a diverse, healthy biological system into one in which only less susceptible organisms can thrive or one devoid of higher organisms.

Prior to planning a mining operation the potential for acid generation is generally estimated based on cores, drill cuttings, and bulk samples used to characterize the orebody. The characterization of acid-generation potential is critical for constructing waste dumps in a way that minimizes the release of contaminants by oxidation of sulfidic rock. For sites where acid generation is a potential problem, characterization of the rock for contaminant release should be conducted concurrent with characterization of the ore potential. Common tests for acid-drainage potential include acid-base accounting and humidity-cell tests. Both

---

[1] Additional information on acid drainage can be obtained from Web-based information sources, including the Acid Drainage Technology Initiative (*www.mt.blm.gov/bdo/adti*), the International Network for Acid Prevention (*www.inap.com.au/inap/homepage.nsf*), and the Mine Environment Neutral Drainage (MEND) program (*mend2000.nrcan.gc.ca*).

methods can underestimate or overestimate acid generation, and continuing research would help to improve the predictions. Predictive tests of acid-rock drainage can be inaccurate if the samples tested do not adequately represent the material that will ultimately be placed on the waste-rock pile or heap. Therefore, careful characterization of the ore and waste rocks is essential.

The preferred technique for managing acid drainage is prevention, which can be effected by several existing methods (Blowes et al., 1995; Lawrence, 1996), including avoidance of acid-generating rock, the addition of inhibitors of microbial oxidation, encapsulation of acid-generating rock in the center of a waste-rock dump that contains mostly neutralizing rock, and the placement of acid-generating rock under water to prevent the rapid transfer of oxygen to reactive surfaces (Steffan et al., 1989). Although each of these techniques has been demonstrated in specialized circumstances, none can be applied in all cases.

The most common method of treating acid drainage is lime precipitation (Kleinmann, 1997). Lime is added to the acidic water and aerated to oxidize soluble ferrous iron to ferric iron, which then precipitates as iron oxide or oxyhydroxide. As the pH is raised, other metal oxides can also be precipitated. The technology for lime treatment of acid drainage, which is reasonably mature and widely utilized, requires the continuous delivery of lime. Unfortunately, the process also generates a calcium-sulfate/iron-oxide sludge that, depending on the concentration of metals and acidity, can pose significant problems. Even if the chemical composition of the sludge is not a concern, this large volume of material can be very difficult to handle.

Other methods of treating acidic drainage, including direct electrowinning of concentrated waste solutions to remove certain metals (an emerging technology), reverse osmosis, and ion-exchange technologies, can be used under certain conditions although at present few of them can compete with lime precipitation. Because acid management is a significant problem in many mines, additional cost-effective methods for treating these fluids are very much needed. For example, if new regulations stipulate that sulfate must be below 500 mg/L, lime precipitation will not be effective. Processes that recover contaminants in a form that can be beneficially used would be particularly attractive.

One promising option that could be used either alone or in combination with lime precipitation is sulfate-reducing bioreactors (Kleinmann and Hedin, 1993; Lawrence, 1998). For this treatment an acclimated biological reactor is fed a variety of alcohols, sugars, other organic substrates, or hydrogen to supply reducing equivalents for the bacteria, which are usually a *Desulfovibrio* species. Sulfate is reduced to hydrogen sulfide, which then precipitates divalent metals as metal sulfides. Like lime precipitation, this treatment creates a sludge (metal sulfides) that is reactive and must be managed to keep it in a chemically reduced state. If the metal sulfides are allowed to reoxidize, the resulting sulfuric acid and metals can again be released. Bioreactors have been used successfully in a variety of pilot plants, as well as for low-volume (1–50 L/min) flows from acidic seeps. The cost of this option can be either competitive with or less expensive than lime precipitation because bioreactors can be operated without electricity, and gravity is used for flow systems.

Passivation technologies that can coat pyrite surfaces to prevent oxidation are also under development. At least three patented methods of passivation are being evaluated. The alkaline-permanganate treatment involves rinsing pyrite surfaces with a potassium permanganate solution at pH >12 (Marshall et al., 1998). This technology was developed by the DuPont Corporation and was recently donated to the University of Nevada. The permanganate oxidizes the pyrite and immediately lays down a mixed iron-oxide/manganese-dioxide coating that is remarkably stable. The silica microencapsulation technology of KEECO utilizes an alkaline solution of silicate to cover the reactive pyrite surfaces (Mitchell et al., 2000). The Envirobond treatment uses a separate proprietary process to form a stable, insoluble compound with pyrite surfaces (Gobla et al., 2000). Phosphate coating uses a soluble form of phosphate that coprecipitates with iron oxide to form a stable surface coating on the pyrite. In silicate coating, basic solutions of silicates are allowed to cover pyrite and form a stable silicate surface on the reactive rock. Further development will be necessary to determine how long any of these methods will last, as well as the costs of treatment and, if necessary, retreatment. The economic viability of these methods is still unproven and will have to be compared to continuous treatments in terms of environmental impact and long-term stability.

Closure and Reclamation of Dump-Leaching and Heap-Leaching Operations and Tailings Impoundments

Upon the cessation of production, dump-leaching and heap-leaching piles and tailings impoundments must be closed in an environmentally sound manner. In the base-metal and precious-metal mining sectors and in some types of industrial-minerals operations these piles and tailings can present potential environmental risks. Depending on the chemical characteristics of the wastes and reagents used, as well as on atmospheric precipitation rates, piles and tailings may contribute poor-quality seepage or runoff to surface and/or groundwater through the release of residual solution or from infiltration of or contact with atmospheric precipitation. The released solution may be acidic (as discussed above) or may contain cyanide or other contaminants, such as selenium, sulfates, radionuclides, or total dissolved solids. Waste piles are also subject to erosion, which potentially contributes sediments directly to surface waters. Over time, erosion may lead to structural instability and potential catastrophic failure. Moreover, waste piles remain as unproductive disturbances on the land unless the land is reclaimed for postmining use.

Mining facilities currently address the closure of dump-leaching and heap-leaching sites in a number of ways. Diversions are used to prevent run-on of precipitation. In the gold industry, heaps are rinsed or fluids recirculated until cyanide in the effluent is reduced to an acceptable level. Spent ore and tailings may be recontoured to limit erosion and enhance long-term structural stability. Cover material may be placed on the surfaces both to inhibit infiltration (and thereby limit leaching) and to provide a medium for plant growth. Alternatively, wastes may be directly vegetated with the addition of soil amendments. In the last decade a number of facilities have run cattle-feeding operations on closed tailings impoundments to create a soil and promote plant growth.

Although existing technologies have been successful, each has limitations. Diversion structures require periodic maintenance to continue to function effectively. Rinsing is typically effective for cyanide, but if contact with air and water is not limited, sulfidic waste materials can continue to oxidize and release contaminants through infiltration of precipitation and subsequent leaching. Covering with soil is a proven technique for revegetation but is only practical if soil is readily available. If no soil was salvaged prior to construction (e.g., at older sites that predated a regulatory requirement of stripping and stockpiling soil for future reclamation), providing a soil cover for a heap or tailings impoundment often requires the creation of an additional disturbance. Moreover, the excavation and placement of large volumes of soil can be costly. A more serious problem, however, is that, regardless of the effectiveness of soil cover as a plant-growth medium, it may not effectively eliminate long-term acid generation and leaching of contaminants.

Ideally, closure technologies for dumps, heaps, and tailings impoundments must be cost effective, require little or no maintenance, limit or reduce the infiltration of precipitation as necessary to control seepage, and must be capable of supporting diverse vegetation. Designs for covers or caps on mine wastes should maximize evapotranspiration in areas of limited precipitation. In areas where precipitation exceeds evaporation cover technologies must seal the surface of mine wastes to prevent the continued oxidation of sulfides and the infiltration of water. Techniques that establish and maintain chemically reducing conditions and thereby immobilize metals should also be investigated. Techniques for the in-situ destruction of cyanide mine wastes should be researched to reduce the time and fresh water required to decommission precious-metals heap-leaching facilities.

Pit Lakes

Open-pit mines that penetrate the water table are "dewatered" to maintain dry working conditions during their operational life. (Underground mines commonly require dewatering as well, but postmining recovery generally returns the groundwater to nearly the same conditions as before mining.) When operations cease and pumping stops, the pit begins to fill (Figure 5-1). The quality of the

FIGURE 5-1 Photograph of pit lake. SOURCE: Glenn Miller, University of Nevada, Reno.

pit water and its potential for adverse impacts on the environment depend on the pit geology and geochemistry, area hydrogeology, and the presence of exit pathways for the pit water. Sulfide materials in the pit walls and floor of base-metal and precious-metal mines begin to oxidize during the mine's operational phase. As the water level rises in the pit and contacts these areas, the acid generated by this oxidation may lower the pH of the pit lake (see above discussion of acid rock drainage). Although the acidification of the pit water can be severe, the rate of oxidation of wall rock decreases once a pit fills with water and the sulfides are no longer exposed to oxygen.

If the acidity and metal levels of the resulting pit lake are not neutralized and buffered by alkaline minerals within the pit, they may pose a threat to the health and safety of humans and wildlife. The rising pit water may encounter other permeable strata or old underground mine workings that may provide conduits for the poor-quality water to enter other aquifers, which can result in discharges to surface drainages.

Existing technologies to address the issue of pit lakes include predictive modeling, in-situ treatment by lime addition, and continued dewatering of the pit to prevent it from filling with poor-quality water. The accuracy of currently available models in estimating the quality of pit lakes is limited by the lack of data necessary to characterize the complex geology and hydrology typically encountered in mineralized regions, as well as uncertainties in predicting the acid-generation potential of sulfide materials (as discussed above with respect to acid rock drainage) (NRC, 1999b). Current models, therefore, can significantly overestimate or underestimate the occurrence or severity of problems. Research to increase the understanding of the complex hydrology of mines may also help to minimize dewatering requirements.

The addition of lime to acidic pit lakes has successfully returned the water to a neutral pH and reduced metal concentrations. This technology is costly, however, especially for large pits with highly acidic water. Continued pumping of groundwater after operations cease eliminates the formation of a poor-quality pit lake but requires a perpetual, not necessarily productive use of energy and groundwater. In addition, it assumes that the resources (e.g., equipment, labor, finances) will remain available indefinitely. Even then, the oxidation of sulfide materials continues as long as the pit remains dry, which eventually results in more severe acidic conditions in the pit lake if pumping is eventually terminated and the pit fills with water.

The industry needs better and less expensive methods to ensure that the acid and metal contents of pit lakes do not threaten humans and wildlife. With more accurate predictive models for final pit lake quality potential problems could be identified during the mine planning process. If the potential for acid generation within the pit is significant, technologies that can reduce or eliminate the oxidation of sulfides in pit walls and floors would be extremely helpful. Cost-effective technologies are also needed for the treatment of acidic and contaminated pit waters at sites that have already experienced this problem.

Treatment of Nonacidic Waters

Some nonacidic aqueous discharges from metal and industrial-mineral mining operations require treatment to remove certain constituents before the water can be discharged into a stream or lake. Dewatering operations frequently generate very large flows that must be treated to meet rigorous discharge standards. The treatment for these high-volume discharges, which contain low concentrations of certain elements, such as arsenic and selenium, is coprecipitation with ferric iron followed by filtration to remove the precipitated elements. Although low levels of arsenic can be effectively removed by this technique, low concentrations of selenium are more difficult to remove by coprecipitation. The removal problem is exacerbated if the solution has other constituents that interfere with the selenium-removal chemistry. Treatment by reverse osmosis, a technology with high capital and operating costs, is sometimes necessary.

Some states also have rigorous discharge standards for nitrate. Mine dewatering products sometimes become contaminated with nitrate as a result of contact with blasting agents, and this water must be treated before discharge. Technology for nitrate removal is limited. Biological nitrate reduction often used in other industries has limitations. Water pumped from mines is often cold, which slows the biological reactions, and is usually devoid of the nutrients required for biological activity. This necessitates the addition of expensive food sources for the bacteria.

Tailings are usually managed by evaporation; however, occasionally overflow from tailings dams must be treated before discharge. In gold mining, cyanide recovery or destruction and metals removal are required. Cyanide recovery, which may be desirable because of the high cost of this reagent, is possible using technology based on a sequence of steps involving acidification, stripping, adsorption, and neutralization. Cyanide and metal-cyanide complexes can be destroyed with hydrogen peroxide, Caro's acid, sulfur dioxide, or by biological means. Following the chemical destruction of cyanide, metals can be precipitated by lime treatment (discussed earlier). When biological treatment is used for cyanide degradation, the resulting nitrate is reduced to gaseous di-nitrogen, and the metals can be removed from the water by binding to the micro-organisms attached to rotating biological contactors.

Better treatment technologies for the effective removal of low concentrations of certain elements, particularly selenium, from large volumes of solution to meet stringent discharge standards, and for efficient cost-effective nitrate removal from aqueous discharges, would be extremely useful.

Slurry Management

Large volumes of water slurries containing fine particles are produced by all types of mining facilities. The management of these slurries as they are dewatered and disposed of can present significant environmental issues. In addition to the potential chemical concerns previously discussed with regard to closure and reclamation, management of these materials can present challenging physical problems.

Whether slurries are produced as tailings from milling operations, spoils in coal mining, or as clay slimes in the phosphate industry, they are often slow and difficult to dewater and dry because of their colloidal nature. The large amount of water tied up with these materials can compromise the structural stability of the units used for their management and disposal, and this situation can be exacerbated by continued precipitation. Although safety factors are routinely included in the engineering design of units that manage these wastes, periodic structural failures continue to occur.

Slime management in the phosphate industry is particularly difficult. Despite years of research few advances have been made in the dewatering and drying of these wastes. When sand is available, it can be mixed with the clay, and the mixture is discharged into impoundments. These sand/clay impoundments eventually dry and become suitable for pastureland, but the sites are not suitable for construction. In addition, mixing clay slimes with sand is a limited option because sand is not always available.

Research to alleviate these problems should promote the rapid dewatering of mine slurries. Research should also focus on ways to ensure that the disposal units for slurried mine wastes remain structurally stable throughout their operational life and beyond. For slimes in the phosphate industry, the development of an economical method of removing water from the clays at the washer circuit and consolidating the clays rapidly would accelerate the return of slimes-management areas to productive use and might increase the options for postmining uses of these areas.

Methane Gas

Methane gas, inherently associated with coal beds, is released as the coal is mined. Methane is a greenhouse gas that contributes to global warming. Some mines capture the gas in methane drainage systems, but these systems on average recover only about 50 percent of the methane evolved in the mines. In mines containing less methane the majority of the methane is released to the mine ventilation system and emitted to the atmosphere in dilute concentrations with the ventilation exhaust (Kruger, 1994).

Research is needed on better methods of collecting methane in drainage systems in the mine, as well as methods of using the low concentrations of methane in the mine return air. In addition to reducing emissions of greenhouse gases advances in these areas would have the health and safety

TABLE 5-1 Opportunities for Research and Technology Development for Environmental Protection

Acid Rock Drainage
- identification of potential acid-generating materials
- prevention of acid generation (encapsulation of wastes, passivation)
- treatment of acidic wastewater

Closure of Dump-Leaching and Heap-Leaching Operations and Tailings Impoundments
- cover technologies to promote evapotranspiration, inhibit infiltration, and/or immobilize metals
- in-situ destruction of cyanide
- alternatives to current production and management of phosphogypsum

Pit Lakes
- improved predictive modeling
- techniques to eliminate or reduce acid generation in pit walls and floors
- treatment technologies for acidic pit water

Treatment of Nonacidic Waters
- removal of low concentrations of metals from large volumes of wastewater
- removal of nitrate from wastewater

Slurry Management
- cost-effective methods of dewatering and consolidating slimes
- methods to ensure the long-term stability of disposal units

Methane
- improved methods of methane drainage collection and recovery of diluted methane from mine ventilation exhaust

Fine Particulates ($PM_{2.5}$)
- control of emissions of fine particulates

benefit of limiting dangerous conditions in mines and the economic benefit of producing a valuable commodity.

Fine Particulate Matter ($PM_{2.5}$)

Federal air-quality regulations limit the concentration of fine particulate matter (2.5 microns or less) in ambient air ($PM_{2.5}$). Although particulates in this size range are emitted from all types of mining operations, industrial-mineral facilities may have the most difficulty meeting this standard partly because they are frequently in urban locations and lack sufficient buffer areas around their mines. The entire mining industry would benefit from control technologies to limit these emissions to acceptable regulatory levels.

RECOMMENDATIONS

The greatest environmental benefits for the mining industry would be derived primarily from advances in the protection of surface and groundwater quality (Table 5-1). Accurate, real-time methods of characterizing the potential of waste materials to generate acid rock drainage would enable operators to identify potential problems before they develop. Research should focus on improving techniques for managing these wastes to prevent acid generation, including biologic and physicochemical passivation methods to inhibit the oxidation of sulfide materials. Research could also focus on further development and optimization of treatment technologies for acid rock drainage, such as biologic reduction, as well as acid rock drainage associated with the pit lakes.

Techniques to improve the long-term environmental stability of closed dump-leaching and heap-leaching operations and tailings impoundments are also areas for additional research. The industry particularly needs improved cover technologies that can enhance evapotranspiration in areas of low precipitation, prevent infiltration of precipitation in areas where recharge to mine-waste materials is significant, and/or maintain reducing conditions in order to immobilize metals. The mining industry may also benefit from research being conducted on impermeable caps and other technologies to prevent the release of toxic materials from hazardous and municipal wastes. Research on the in-situ destruction of cyanide could lead to quicker, less water-consumptive methods for closures of precious-metals heap-leaching facilities.

Alternatives to the current generation and management of phosphogypsum would also be beneficial.

Although much of the recommended research involves acid rock drainage, other areas for fruitful research are improving technologies for managing nonacidic wastewaters, including effective, low-cost techniques for removing low concentrations of elements, such as selenium, from large volume flows and for removing nitrates from wastewater discharges. The dewatering of phosphate slimes and other slurried mine wastes, as well as the long-term stability of disposal units for these wastes, could also be an area for future research. Development of better techniques for methane recovery from underground coal mines would provide environmental, health and safety, and economic benefits. Research is also needed on technologies to control emissions of fine particulates.

In many respects the technology needs identified in this chapter either represent incremental advances or practical developments of known technologies or build on techniques already known and understood to some degree. The committee believes that research into areas beyond our current understanding, and which may now appear to be infeasible, could lead to innovative solutions to environmental problems facing the mining industry (Sidebar 5-2).

SIDEBAR 5-2
Blue Sky Ideas for Research on Environmental Issues

- **Economical recovery of metallic constituents from acid rock drainage releases, including methods of recovering iron sulfides in a form that could be used as a soil acidifier in agriculture**. Technologies that could turn costly waste-treatment projects into economically viable production activities would have significant benefits in terms of resource conservation and waste minimization, in addition to the clear economic benefit of creating a secure and dedicated funding source for the remediation of acid rock drainage.

- **Sealing of pit walls to prevent the oxidation of sulfides and the formation of acidic pit lakes**. Open-pit mine walls at or above the water level of a pit lake can continue to oxidize and adversely affect the quality of pit water if they contain unbuffered sulfide minerals. Significant technical barriers must be overcome to devise a cost-effective way to seal these rock faces and prevent their long-term oxidation.

- **Sealing of in-situ mineralized zones to prevent the inflow of groundwater and allow in-situ leaching and complete recovery of solutions without continuous pumping to maintain a cone of depression**. The migration of metal-rich leaching solutions beyond the mineralized zone of an in-situ orebody and into the surrounding aquifer is currently controlled by continuous pumping of the groundwater to create and maintain a cone of depression around the operation. The cost of this dewatering activity, as well as the use of the water resource, could be eliminated if the mineralized zone could be sealed in place. Better control of the leaching solution would also provide a significant environmental benefit.

- **Remote sensing of groundwater quality throughout the operating life of a mine and during the closure and postclosure periods**. Groundwater-monitoring programs at mine sites can involve large numbers of wells, require thousands of samples and analyses, and last for decades. Techniques for direct, down-hole measurements and data logging could significantly reduce the cost and human resources requirements of monitoring.

- **Effective, inexpensive technology for removing low concentrations of contaminants (e.g., selenium, arsenic) from high-flow volumes of water**. The cost-effective recovery of low concentrations of potentially toxic large flows would have significant economic and environmental benefits.

- **Techniques for rapidly developing soils on mine wastes, heaps, and tailings**. The direct vegetation of mine wastes, without a soil cover or amendments designed to create a suitable growth medium, may foster some plant growth but can result in the establishment of a monoculture that cannot support complex ecosystems. Research should be conducted to develop cost-effective techniques for creating soil that can support a diverse plant community directly from mine waste.

6

Current Activities in Federal Agencies

Since the demise of the USBM in 1996, the level of federal research to assist the mining industry in the United States has fallen and has not been well coordinated. One exception is research on occupational health and safety in the mining industry, which is being overseen by NIOSH. As awareness of environmental concerns has increased, numerous research programs have been initiated throughout the federal government to develop new and improved methods of remediating historic metal-mine and coal-mine wastes. Information on mineral production in the United States and outside the country is available from the Bureau of Census in the U.S. Department of Commerce, the USGS (often in cooperation with state geological surveys), and the DOE Energy Information Administration, although data collection is not well coordinated among these agencies.

One area that is no longer overseen by a federal agency is research to improve the efficiency and effectiveness of mining technologies. However, many other federal research and development programs dealing with transportation, excavation, basic chemical processes, and novel materials could ultimately be of help to the mining industry and the nation. The only active federal program dealing solely with the development of more efficient and more environmentally benign mining technologies is the Mining Industries of the Future Program, a component of the IOF Program of the DOE's OIT. A number of federal agencies are also involved in science, engineering, and technology development that could be useful for the mining sector (Appendix C).

U.S. DEPARTMENT OF AGRICULTURE

The U.S. Department of Agriculture is responsible for the national forests, which contain a number of coal, metallic, and industrial-mineral mining districts. The U.S. Forest Service assists in mineral exploration by maintaining a minerals and geology inventory for national forest lands. The minerals and geology management group within the agency oversees the restoration and reclamation of the land and watersheds affected by historic mining practices. The Forest Service is part of the federal Interdepartmental Abandoned Mine Lands Watershed Cleanup Initiative, which also involves the Bureau of Land Management, National Park Service, USGS, and Environmental Protection Agency. Together these agencies are developing a coordinated strategy for the cleanup of environmental contamination from abandoned hardrock mine sites on federal lands. Although the initiative does not directly support research on remediation technologies, various agencies, including the U.S. Department of Agriculture, have individual programs that support in-house and grant-supported research.

U.S. DEPARTMENT OF COMMERCE

The National Institute of Standards and Technology (NIST) conducts the Advanced Technology Program, which supports a wide variety of cost-shared projects involving the government and the private sector. The majority of projects are focused on information technology, biotechnology, and materials research. Current projects that could have immediate spin-off applications for the mining industry include intelligent control, membrane and other separation technologies, microsystem and nanosystem technologies, and catalysis and biocatalysis technologies. No NIST programs are directly focused on the mining industry, although the agency recently conducted a study of mine fires.

U.S. DEPARTMENT OF ENERGY

Currently, most of the federal engineering and technology development that is focused on or could be useful to the mining industry is being conducted by DOE. Programs in many parts of the department, but especially in the national laboratories (Table 6-1), have yielded advances relevant to the mining industry. The OIT Mining Industries of the Future Program is currently the only agency program focused solely on improving the energy efficiency, resource utiliza-

TABLE 6-1 Estimates of Mining Research and Development Capabilities of the National Laboratories

	Albany Research Center	Ames Laboratory	Argonne National Laboratory	Brookhaven National Laboratory	Idaho National Engineering & Environmental Laboratory	Kansas City Plant	Lawrence Berkeley National Laboratory	Lawrence Livermore National Laboratory	Los Alamos National Laboratory	National Energy Technology Laboratory	National Renewable Energy Laboratory	Oak Ridge National Laboratory	Pacific Northwest National Laboratory	Sandia National Laboratories	Savannah River Technology Center	Y-12 Plant
Exploration and Resource Characterization																
Exploration optimization strategies	C				C		S	S	C			C	C	S		
Remote sensing for geophysical systems		S			SC	C	S	S	S	S		C	C	S	S	
Subsurface imaging and characterization/ Geologic basin modeling			SC	S	SC	SC	S	S	S	SC		C	S	S	S	
Explosive design and engineering									S					S		
Satellite imaging/data interpolation	C		C		SC	C	C	C	S			SC	S	S		
Rock mechanics						C	S	S	C			C		S		
Advanced drilling technologies					C		SC	SC	S	C				S		C
Geochemical sensors and analytical devices		C	SC	C	S		S	S	S	SC	C	SC	SC	S	SC	C
Sensors and analytic devices for imaging physical material properties	C	C	S	S	S	C	S	S	S	C	C	C	S	S	S	SC
Mining Operation																
Gas, aerosol, noise sensors			SC	SC	SC		SC	SC	SC			S	C	S	S	C
Robotics/automation for autonomous equipment				S	C		SC	SC	SC			S	SC	S	S	C
Fuel cells/power supplies		C	S	C	C		S	SC	S	S	SC	C	SC	SC	SC	C
Alternative fuels/propulsion systems			S	C	SC		S	SC	C		S	C	C	S	SC	SC
Information and data systems in support of autonomous technologies			S		SC	S	SC	SC	S			C	SC	C	SC	SC
Mine climate simulation									S				C	C		
Advanced materials (e.g. strength, durability, hardness)	S	S	S	C	S		S	S	S	C		S	S	S	SC	C
Minerals Processing																
Separations technologies for solids processing	SC		S		SC		C	C		S		C	SC	SC	SC	
Dewatering and water-reuse technologies	SC		SC		SC		C	C		S		S	S	SC		C
Biological processes for in-situ metal extraction				S	S		SC	C		SC	C	SC	SC			
Geochemistry for minerals processing	SC				S		SC	S	C	C		SC	C	S		
Real-time analytic techniques and models		SC	C	C	S	C	SC	C	S	S	SC	SC	S	S	S	SC
Alternative technologies for processing by-products and alternative uses for products	SC		SC	SC	SC		C			S		C	SC	SC		
Sensors for process-stream characterizations and control		SC	SC		S		SC	SC	S	C	SC	SC	S	S	S	SC

S – strength; SC – significant capability; C – capability.
DOE defines a strength as an area with 25-30 or more full-time-equivalent positions in the past five years; a significant capability as 10-24 full-time-equivalent positions in the last 5 years; and a capability as 5-9.9 full-time-equivalent positions in the last 5 years.
SOURCE: OIT, 2000.

tion, and competitiveness of the mining industry. The purpose of the program, a collaborative partnership with the NMA, is to demonstrate, evaluate, and accelerate the development of new technologies and provide scientific insights into the needs of the mining industry. DOE has a cost-shared cooperative agreement program in place to support innovative, precompetitive projects that demonstrate potential energy, environmental, and economic benefits for coal, metal, and industrial-minerals mining. Projects are 50 percent cost shared with the government and must have at least a 10 percent contribution from industry. The program has already awarded a second round of industry partnerships and is soliciting proposals for a third round in the area of mineral processing.

The DOE Office of Transportation Technologies supports research into technologies to reduce emissions from heavy vehicles and to develop advanced automotive materials, advanced batteries, and alternative fuels. Although none of these programs is focused specifically on the mining industry, research on new technologies for heavy vehicles could have broad implications for future mining methods. Alternative vehicle technologies for mining are also being explored by the Hydrogen Program in the DOE Office of Power Technologies, which is funding the development of a fuel-cell-powered mining vehicle. Drilling technologies are also being explored by the Office of Power Technologies through grant awards to universities and the private sector. Research supervised by the Geothermal Drilling Organization focused on the development of new and improved drilling technologies for geothermal fields, such as slimhole drilling and improved instrumentation, is also of interest to the mining industry. Research on drilling is also being funded by the DOE Office of Fossil Energy. This research is primarily focused on oil and gas drilling but has a large overlap with excavation drilling applications. The Office of Coal and Power in the DOE Office of Fossil Energy also supports research on coal processing for various applications.

Engineering development on many fronts is being conducted in DOE national laboratories. Work at the Albany Research Center, a former USBM laboratory, is focused on mineral processing and metallurgy. The National Renewable Energy Laboratory is investigating coal processing as part of its mission to investigate alternative technologies for processing by-products. Mining technologies, particularly transportation, robotics, sensors, and instrumentation, are a focus of research at Lawrence Berkeley Laboratory and the Idaho National Engineering and Environmental Laboratory, which also has a fractured-rock science team whose task is to integrate interdisciplinary research activities at other national laboratories, federal agencies, and universities. Research and technology development at the national laboratories could have important implications for the development of new mining technologies. For example, focused research is being done at Argonne National Laboratory on the development of analytic tools and sensor technology, which would be especially useful for resource recovery. The Los Alamos National Laboratory has a long-standing interest in drilling technologies through its support of geothermal power-generation technologies. Sandia National Laboratory has focused research on fuel-cell technologies for transportation that could impact mine transportation; this laboratory is also investigating various means of underground imaging.

Research into technologies for the remediation of waste materials is being undertaken at a number of the national laboratories, as well as through the DOE Office of Environmental Management. Many of the techniques being developed may have applications in the mining industry. The Mine Waste Technology Program operated jointly by the Idaho National Engineering and Environmental Laboratory and the Environmental Protection Agency is focused on metallic mining wastes. The Office of Basic Energy Sciences supports research on basic chemical processes.

Although the predecessors of DOE, along with the U.S. Department of the Interior, at one time conducted research related to the exploration and development of nuclear fuel resources (uranium and thorium), almost no research is being conducted by federal agencies in these areas today.

U.S. DEPARTMENT OF DEFENSE

Although no DOD research is focused directly on the mining industry, a number of research projects could have spin-off benefits for the industry. The geotechnical laboratory at the U.S. Army Corps of Engineers Waterways Experiment Station specializes in soil and rock mechanics, slope stability, seepage analysis, engineering geology, geophysics, dust control, and vehicle mobility. Although this research is focused on military applications, much of the data produced would also be applicable to the mining industry.

The Army Excavation Logistics Center and the Defense Threat Reduction Agency both support research into excavation and penetration that could be of direct benefit to the mining industry. The Defense Advanced Research Projects Agency (DARPA) funds innovative concept research in a wide variety of fields that may be applicable for national defense. Although, few existing projects appear to be directly applicable to the mining industry, investigations of novel vehicle designs by both individual DARPA investigators and the Office of Naval Research may provide interesting new concepts for mining. The Army Research Laboratory conducts research on robotics, excavation technologies, sensors, and materials, all of which could have direct mining applications. The Office of Naval Research, the Naval Research Laboratory, the Air Force Office of Scientific Research, and the Air Force Research Laboratory are developing sophisticated communications technologies and means of underground imaging. The Air Force Space Systems Command is evaluating hyperspectral equipment (SEBASS), which could be useful for mineral exploration. This system is a midwave, long-range imaging spectrometer for remote sensing that operates from an airborne platform.

Like many other agencies, DOD is conducting research into environmental remediation at its own laboratories and through grants and fellowships that may impact mine cleanup technologies.

U.S. DEPARTMENT OF HEALTH AND HUMAN SERVICES

NIOSH conducts research on reducing miners' occupational injuries and illnesses, primarily at its Pittsburgh and Spokane research centers. Although most of this research is conducted in house, NIOSH also works with the coal- and metal-mining industries to develop test beds and new technologies and provides grants to universities for applied research. Areas of active research include the design of equipment to prevent injuries; control of diesel emissions and emissions of toxic substances; control of dust and silica; monitoring of dust; development of emerging technologies; prevention of fires and explosions; improved ventilation; ground control of coal; ground control of metals and nonmetals; hazard detection; prevention of hearing loss; understanding of human factors; surveillance and statistical activities; and training and education.

U.S. DEPARTMENT OF THE INTERIOR

As part of its mandate to manage and protect large portions of the federal land system, the U.S. Department of the Interior is involved in mining issues. The Bureau of Land Management and the Office of Surface Mining Reclamation and Enforcement provide regulatory control for existing mining operations. The Office of Surface Mining is particularly active in the transfer of information technologies applicable to regulating the mining industry to local, state, and federal governments. The agency provides technical assistance to coal-mining programs.

A number of groups are working on the remediation and reclamation of abandoned mine sites. These groups include the Bureau of Land Management, the Office of Surface Mining, the National Park Service, the USGS, and the Bureau of Reclamation. The USGS is particularly active in the study of acid mine drainage and the development of technologies to remediate historic sites. The Office of Surface Mining supports the Abandoned Mine Land Program, which repairs, reclaims, and restores as much land and water as possible through a fund supported by fees collected from active coal-mining operations. As part of the Surface Mining Control and Reclamation Act of 1977, the federal government has collected more than $5.5 billion in taxes on coal production. Congress has allocated a large portion of these funds to states and tribes, which have leveraged the monies by forming partnerships with federal and private interests and have reclaimed more than 5,500 problem areas and more than 180,000 acres, using both traditional and innovative technologies.

The USGS supports mineral-exploration research for the coal, industrial-minerals, and metal-mining industries. The basic level of support is the development of geologic maps by the Geologic Division, which identifies potential areas with mineral resources. The Mineral Resources Program focuses on land stewardship, national and international commodity studies, mineral conservation, and materials flow. In addition, the program develops geologic ore-deposit models through detailed investigations of specific ore deposits and ore districts, which can be helpful in mineral exploration. Geophysicists in the program are working on the development and application of new geophysical technologies to the search for mineral deposits. Remote-sensing technologies and applications are also being developed by the USGS. These technologies are applicable to the exploration for and identification of wastes from the mining industry.

The Division of Energy and Mineral Resources at the Bureau of Indian Affairs helps tribes identify and promote the development of mineral resources on their reservations. As part of this program the Bureau has developed the National Indian Energy and Mineral Resources Data Base to collect and archive all data and information nationwide. The program also conducts limited mineral exploration and has worked on the development of exploration models.

U.S. DEPARTMENT OF LABOR

The MSHA enforces laws on mine health and safety and provides technical support to its enforcement arm and the mining industry. The technical support team at MSHA, primarily at the Pittsburgh Safety and Health Technology Center, conducts field investigations, laboratory studies, and analyses to resolve specific problems related to dust and other physical and toxic agents, mine electrical systems, mine emergency operations, mine waste, geotechnical engineering, roof control, and ventilation. MSHA also evaluates equipment and materials used in mines, primarily through the Approval and Certification Center. MSHA's National Mine Health and Safety Academy develops resources and offers training programs for miners and mine supervisors.

U.S. DEPARTMENT OF TRANSPORTATION

Although the U.S. Department of Transportation is not directly involved in research focused on the mining industry, research supported by the Federal Railway Administration on new railroad vehicle and safety technologies and research by the Federal Transit Administration on vehicular design and propulsion technologies could have important spin-offs for the mining industry. The Department of Transportation also funds research on the use of aggregate and sand in roadways through the Federal Highway Program.

U.S. ENVIRONMENTAL PROTECTION AGENCY

The U.S. Environmental Protection Agency funds in-house and grant-supported research by academic and industry groups on remediation and reclamation technologies. Programs dealing with abandoned mine lands are of primary interest to the mining industry. These programs include treatment technologies ("end-of-the-pipe"), source-control technologies, and to a lesser extent resource-recovery technologies. Active research areas in source-control technologies include sulfate-reducing bacteria technologies; biological destruction of cyanide for heap-leaching operations; transportation-control and pathway-interruption techniques, including infiltration controls, sealing, grouting, and plugging by microbiological systems; and acid mine-drainage production and prediction techniques. Because mine wastes are volumetrically large at Superfund sites, remediation is the focus of many projects in the Superfund Innovative Technology Evaluation (SITE) Program. Results of these research programs could significantly advance the design of more environmentally benign mining operations, especially in-situ mining. Data from scoping studies of proposed mining operations, such as the Copper Range Company Solution Mining Project (White Pine, Michigan), or remedial actions could also provide the mining industry with important information for the development of innovative technologies.

The Mine Waste Technology Program, operated by the Environmental Protection Agency with industry and university partners and support from the DOE Environmental Management Program, supports the development and testing of technologies for long-term and short-term solutions to mine waste remediation and resource recovery.

NATIONAL AERONAUTICS AND SPACE ADMINISTRATION

Research at NASA impacts the mining industry primarily through the development of tools and technologies that would be useful for mineral exploration. Remote sensing technologies developed by NASA have transformed mineral exploration, and the development of increasingly sophisticated sensors, such as the AVIRIS and the AIRSAR, promise further advances in the direct detection of mineral resources. Research into planetary geology and basic Earth sciences by the agency and in cooperation with universities also provides important information for mineral exploration. New drilling technologies being developed to support the exploration of Mars will also have significant implications for the mining industry. The results of drilling research being supported by the Jet Propulsion Laboratory will be of special interest to the mining industry.

NATIONAL SCIENCE FOUNDATION

The NSF (National Science Foundation) supports university research in engineering and Earth-science disciplines with direct benefits for the mining industry. The Directorate for Engineering supports research in civil and mechanical engineering, as well as in design and industrial innovation. The Division of Civil Engineering within the Engineering Directorate focuses on tunneling and excavation research. The NSF-supported Engineering Research Center for Particle Size and Technology at the University of Florida is engaged in research of direct interest to the mining industry, particularly the industrial-minerals sector. Research areas at the center include advanced separation processes and technologies, dispersion, agglomeration and consolidation of particles, and engineered particulates. The industry may soon be able to participate directly in other NSF engineering programs that will impact mining technologies through a program for establishing academic liaisons with industry. The NSF Division of Earth Sciences supports fundamental mineralogical, physical, and chemical research on the nature, origin, and temporal evolution of the Earth's crust and research into basic chemical processes. Specific research projects focus on the genesis of mineral deposits, as well as geochemistry and biogeochemistry for mineral exploration, extraction, and remediation.

NONFEDERAL PROGRAMS

Several states have research programs in mining-related technologies. Colorado, Minnesota, Montana, Nevada, and West Virginia support research on mine-waste remediation technologies. In Minnesota the Iron Ore Cooperative Research Program uses industry and state monies to fund research and development projects of interest to the iron-ore industry. The Minnesota Department of Trade and Economic Development provides Taconite Mining Grants to assist in the implementation of technology improvements. In Colorado, research by the Colorado Geological Survey is funded jointly by the state and industry through a levy on mineral producers. In Florida, severance taxes paid by the phosphate industry are used to fund research and development at the Florida Institute of Phosphate Research. The focus of the Minerals Research Laboratory, located at North Carolina State University, is on the beneficiation of industrial minerals.

The National Aggregate Association, in conjunction with the National Stone Association, supports aggregate research at the International Center for Aggregates Research, a joint project at the University of Texas at Austin and Texas A&M. The National Ready-Mixed Concrete Association funds the Cement and Aggregate Research Laboratory at the University of Maryland, College Park. The mining industry supports research on exploration, mining, and minerals processing through individual grants to a number of universities throughout the United States.

RECOMMENDATIONS

Federal activities that could directly benefit the mining industries include the engineering research groups at DOE national laboratories. The U.S. Army Corps of Engineers Waterways Experiment Station Laboratory also focuses on research applicable to needs of the industry. The problem is not the lack of skilled researchers in federal agencies but the lack of projects focused directly on the problems of most interest to the industry.

The federal government has an appropriate role to play in supporting research that directly or indirectly can promote the development of technologies for the mining industry. The committee recommends that the federal government support research and development on mining technology for all stages of the mining process from exploration through mining to mineral processing. The government should have a particularly strong interest in long-term, "blue sky" research that is currently not being undertaken by the industry itself. Although some research applicable to several aspects of mining is currently being conducted by various federal programs, the DOE national laboratories appear to represent one of the most promising venues for undertaking such research.

7

Government-Sponsored Research and Development in Mining Technology

Mining is important to the United States because the country is both a major consumer and a major producer of mineral commodities. This section first discusses the benefits of research and development on new technologies in exploration, mining, and mineral processing for consumers and producers. This is followed by a discussion of the role of government in supporting and fostering research and development in this sector.

BENEFITS OF RESEARCH AND DEVELOPMENT

Allocating Federal Funds for Science and Technology (NRC, 1995c) made a strong case for federally funded research and development. Among other things the study committee noted:

> The federal government has played a pivotal role in developing the world's most successful system of research and development. Over the past five decades the U.S. scientific and technical enterprise has expanded dramatically, and the federal investments in it have produced enormous benefits for the nation's economy, national defense, health and social well being.

Successful research and development produces new technologies that reduce production costs; enhance the quality of existing mineral commodities; reduce adverse environmental, health, and safety impacts; and create or make available entirely new mineral commodities. The resulting benefits may accrue to consumers or producers and communities near the mine operation. Most often, however, both consumers and producers benefit, with consumers enjoying most of the benefits over the long run.

Clearly, producers can and do benefit from new technologies. However, some companies enjoy reductions in costs and increases in profits at the expense of other firms that benefit less or not at all from the new technology. The latter firms may even lose competitiveness and profits. Nevertheless, most of the benefits generated by a new technology normally flows to consumers. This is particularly true in the long run because, as technologies become more widely available, commodity prices drop, and therefore companies and countries can only maintain a technology gap by continually generating innovations that provide new benefits.

ROLE OF GOVERNMENT

Even if the nation as a whole and private firms reap the benefits of new technologies, the appropriate role of industry and government in research and development for exploration, mining, and mineral processing is still an open question. Mining companies that benefit from research and development presumably have an incentive to pay some of the costs. In economic theory one would expect a particular company to be willing to increase its research and development outlays until the value of expected benefits equals the dollars spent. The expected benefits, of course, are not necessarily the same as the actual benefits. Some successful research and development projects produce benefits far beyond expectations; many others fail to reach expectations. The incentive for firms to expand research and development will end when the expected benefits are less than the expected cost. Risk is a major consideration because most of the research and development projects proposed can be expected to fail.

In determining the value of research and development, private firms consider only the benefits they expect to capture. Therefore, the welfare of society as a whole is best served when research and development expand to the point at which the expected benefits of additional expenditures just equal the costs, appropriately discounted for time and risk. The expected benefits include internal benefits to the firm carrying out the research and development, as well as external benefits that consumers and other producers will realize. Because the external benefits in exploration, mining, and mineral processing often constitute a large portion of the total benefits, the market will not support the optimal amount of research and development, possibly by a wide margin, without government support.

> **SIDEBAR 7-1**
> **Benefits of SXEW to Producers and Consumers**
>
> Solvent-extraction/electrowinning (SXEW) is partly responsible for the decline in the real price of copper in the past 30 years. Millions of people who use copper-containing products have benefited. The United States, the world's largest copper-consuming country, has benefited greatly.
>
> SXEW has also improved the competitiveness and profitability of many companies by reducing production costs and enabling them to exploit copper minerals that might otherwise be wasted. The technology was developed and first used by smaller copper mining companies in the United States. Later, SXEW was adapted by foreign producers and major U.S. copper companies.
>
> With globalization and the rapid diffusion of new technologies around the world, conventional wisdom suggests that innovating firms will only be able to maintain short-term competitive advantages. This argument rests on two critical assumptions. First, the innovation is a one-time event, rather than a series of associated innovations over an extended period of time. Second, the new development is neutral in that it affects all producers equally. More often than not, one or both of these assumptions is incorrect.
>
> Today, for example, the SXEW process is the result of first a single innovation in chemical reagents that was followed by literally hundreds of innovations in equipment, instrumentation, and operations over a period of 40 years. These innovations by vendors, operators, and researchers have reduced costs and greatly extended the range of applications of SXEW. As a result, firms and countries can benefit from lower costs for as long as they remain technological leaders.
>
> In addition, SXEW is not a neutral technology. It has particularly helped countries with a history of copper mining that has left many large oxide waste dumps; countries with an existing copper mining industry that must reduce or capture sulfur emissions from its smelters (SXEW requires large quantities of sulfuric acid); countries with arid climates because heavy precipitation can make leaching difficult; and countries with copper mines that produce few valuable by-products because SXEW cannot yet recover valuable by-products economically. In short, SXEW helps the United States and a few other copper-producing countries more than it helps others. This is not entirely surprising because U.S. producers and suppliers have played an important role in the development of the technology.

External benefits, of course, are not unique to advances in the mining industry. Over the research and development cycle, the ratio of external to total benefits tends to fall. For this reason, government funding supports a higher proportion of basic research than development. Activities other than research and development can also create external benefits. Governments around the world, for example, fund education on the grounds that the social benefits of education will far exceed the private benefits.

External benefits are the most important but not the only rationale for government support for research and development. Without detailing the full litany of possible reasons for market failure, we note that the discount rates firms use to assess investment projects, including research and development projects, may exceed the appropriate rates for society as a whole. Therefore, without government support, the private sector tends to underfund research and development, particularly projects that are highly risky and projects that are expected to produce benefits in the distant future. For this reason, the government may have a strong interest in supporting so-called high-risk, "far-out," "off-the-path," "blue-sky" research.

RESEARCH AND DEVELOPMENT IN MINING TECHNOLOGY

Minerals are basic to our way of living. Essentially everything we use is a product of the mining, agriculture, or oil and gas industries, including the things that comprise and operate the tools of the so-called computer-information and communications age. Basic industries (i.e., mining, agriculture, and energy) play a crucial role in our economy.

Because of high capital requirements, small profit margins, the cyclic nature of commodity prices, long lead times for the development of new properties, and environmental constraints, the mining industry historically has been very conservative in initiating and adopting new technologies. Nevertheless, the industry has made significant advances in productivity, environmental control, and worker health and safety.

The development of the SXEW process, for example, has led to the low-cost production of copper from waste and raw-ore dumps of copper minerals (primarily oxides and silicates) found at many copper mines. SXEW is a hydrometallurgical process that differs radically from the traditional method of producing copper by milling, smelting, and refining. Currently SXEW accounts for about a third of U.S. copper production (Sidebar 7-1)

For more than a century, the federal government has been involved in research and development to meet the basic needs of society—food and other agricultural products, energy, and minerals. The role of the government has been largely to foster research that improves efficiency, and therefore reduces costs to the consumers, and lessens the environmental consequences of agricultural, energy, and mineral production. Agriculture research by the U.S. Department of Agriculture and others has contributed greatly to the productivity of farms, the quality of food, and the safety of agricultural products. Energy research and development by DOE and its predecessors has provided better ways to produce and use scarce, clean fuels and has improved our energy security.

The USBM was the focal point for federal research in mineral technology from its inception in 1910 to its demise in 1996. Its accomplishments and contributions to the U.S. economy were significant. For example, the bureau developed fundamental technology for extracting refractory gold, which helped to establish Nevada's modern gold industry. Early work in taconite extraction and processing was very helpful to Minnesota's iron industry. The USBM was also the developer and publisher of thermodynamic data on important mineral-processing systems.

During World War II the USBM developed zirconium and titanium metallurgy. Bureau experts were world leaders in mine health and safety technology. Technology developed by the bureau and its funded research clients in explosives, combustion, and ventilation not only improved mine health and safety but is also widely used in other industrial, civil, and military applications. In addition, the federal government's research and development programs have traditionally funded work in mining schools and other centers of excellence, thereby contributing to the education of engineers and technologists needed by industry and federal and state government agencies.

In the 1980s most facets of the mining industry suffered a major recession. Almost every major mining industry research and development facility was closed, and most companies curtailed their formal research and development programs. In the 1980s and 1990s the equipment companies picked up some of the slack, and the USBM continued to provide a modest funding base for mining schools.

RECOMMENDATIONS

The federal government's current efforts in mineral technology are very small and unfocused. The mining industry continues to progress technologically, but many universities are finding it difficult to obtain funding for mining-specific research. Unless more federally funded research and development programs, such as the IOF Program, are forthcoming, the technological progress of the industry may slow down and eventually affect the education of trained technical people for industry and government. Technology transfer to mining from other industries (e.g., medical, manufacturing, chemical, telecommunications) must also be improved.

DOE's stated objective of energy conservation is achievable in the mining industry, which is a major energy user. Increases in productivity in mining, a reduction in the number and complexity of process steps, and improvements in comminution are all examples of advances that could reduce energy consumption. Improvements in efficiency could further optimize by-product and emission management and supplement the ongoing health and safety research and development done by NIOSH. In addition, more federal participation in mining technology research would meet one of the criteria for government support of research and development: "development of new enabling, or broadly applicable, technologies for which government is the only funder available" (NRC, 1995c).

8

Summary of Conclusions and Recommendations

The Committee on Technologies for the Mining Industries has (1) reviewed information concerning the U.S. mining industry; (2) identified critical research and development needs related to the exploration, mining, and processing of coal, industrial minerals, and metals; and (3) examined the federal contribution to research and development in the mining process. The committee has attended presentations and received information from representatives of government programs, industry, and academia. The committee also reviewed government documents, pertinent NRC reports, other technical reports, and published literature.

The previous chapters have discussed the importance of the mining industry to consumers and the U.S. economy (Chapter 2), presented technological gaps and research needs in exploration, mining, and processing (Chapter 3), and outlined the technological needs associated with health and safety (Chapter 4) and environmental issues (Chapter 5) in the minerals industry. The report has detailed the involvement of federal agencies in mining (Chapter 6) and the importance of federal involvement in mining research and development (Chapter 7).

IMPORTANCE OF MINING TO THE U.S. ECONOMY

Finding. Mining produces three types of mineral commodities (metals, industrial minerals, and fuels) that all countries find essential for maintaining and improving their standards of living. Mining provides critical needs in times of war or national emergency. The United States is both a major consumer and a major producer of mineral commodities, and the U.S. economy could not function without minerals and the products made from them. In states and regions where mining is concentrated, this industry plays an important role in the local economy.

TECHNOLOGIES IN EXPLORATION, MINING, AND PROCESSING

Mining involves a full life cycle, from exploration through production to closure with provisions for potential postmining land use. The development of new technologies has benefits for the mineral industries throughout this full life cycle and for every major component of the mineral industries: exploration, mining (physical extraction of the material from the Earth), processing, associated health and safety issues, and environmental issues. The committee recommends that research and development be focused on technology areas critical for exploration, mining, in-situ mining, processing, health and safety, and environmental protection (Table 8-1).

The mining industries are constantly undergoing incremental or evolutionary changes as uses are found for new technologies developed for other applications. Occasionally, revolutionary changes occur when new technologies, developed either inside or outside the industry, take hold. Chapter 3 provides examples of past evolutionary and revolutionary changes. Given the current rapid pace of technological development in broad areas (from information technology to microbiology), the committee envisions that mining industries will be able to take advantage of some of these developments to the benefit of consumers, producers, workers, and the environment. Progress towards these revolutionary changes will produce concrete developments for industry. These revolutionary changes can result from basic research, applied research, or technology development (Sidebar 8-1).

HEALTH AND SAFETY RISKS AND BENEFITS

Finding. Advances in technology have greatly enhanced the health and safety of miners. However, potential health hazards arising from the introduction of new technologies, which may not become evident immediately, must be addressed as soon as they are identified.

TABLE 8-1 Key Research and Development Needs for the Mining Industries

Research and Development Needs	Exploration, Chapter 3[a]	Mining, Chapter 3[a]	In-Situ, Chapter 3[a]	Processing, Chapter 3[a]	Health & Safety, Chapter 4[a]	Environmental Protection, Chapter 5[a]
Basic Research						
Basic chemistry – thermodynamic and kinetic data, electrochemistry	X		X	X		X
Fracture processes – physics of fracturing, mineralogical complexities, etc.		X	X	X		
Geological, geohydrological, geochemical, and environmental models of ore deposits	X	X	X			X
Biomedical, biochemical, and biophysical Sciences	X	X	X	X	X	X
Applied Research						
Characterization – geology (including geologic maps), hydrology, process mineralogy, rock properties, soils, cross-borehole techniques, etc.	X	X	X	X	X	X
Fracture processes – drilling, blasting, excavation, comminution (including rock-fracturing and rubblization techniques for in-situ leaching and borehole mining)	X	X	X	X		
Modeling and visualization – virtual reality for training, engineering systems, fluid flow	X	X	X	X	X	X
Development of new chemical reagents and microbiological agents for mining-related applications (such as flotation, dissolution of minerals, grinding, classification, and dewatering)			X	X		
Biomedical, biochemical, and biophysical sciences			X	X	X	X
Water treatment						X
Closure					X	X
Alternatives to phosphogypsum production and management						X
Technology Development						
Sensors – analytical (chemical and mineralogical; hand-held and down-hole), geophysical (including airplane drones, shallow seismic data, and hyperspectral data), surface features, personal health and safety, etc.	X	X	X	X	X	X
Communications and monitoring		X		X	X	X
Autonomous mining		X			X	
Total resource recovery without environmental impact		X	X	X		X
Fine and ultrafine mineral recovery (including solid-liquid separation, recovery of ultrafine particles, disposal)				X	X	X
In-situ technologies for low-permeability ores (includes some of the technologies under fracture processes as well as directional drilling, drilling efficiencies, casing for greater depths)	X		X	X		
Biomining		X	X	X		
Fracture processes – applications of petroleum and geothermal drilling technologies to mining	X	X	X			

[a]Justification for including these research and development needs is found in the chapters indicated.

RESEARCH OPPORTUNITIES IN ENVIRONMENTAL TECHNOLOGIES

Finding. The need for a better understanding of the scientific underpinnings of the environmental issues and for more effective technologies to address them cannot be overemphasized.

Recommendation. Technologies that attempt to predict, prevent, mitigate, or treat environmental problems will be increasingly important to the economic viability of the mining industry. Improved environmental technologies related to mine closures present the greatest opportunity for increasing productivity and saving energy. Research is also needed on water-quality issues related to mine closures, which are often challenging and costly to address for all types of mining.

ROLE OF THE FEDERAL GOVERNMENT

The committee agrees with previous NRC studies (e.g., NRC, 1995c) indicating that the federal government has important roles in nearly all areas of basic and applied research and in fundamental technology development (Sidebar 8-2).

> **SIDEBAR 8-1**
> **Potential Revolutionary Developments for Mining**
>
> - In-situ mining of a broader range of commodities
> - Biomining (using biological agents to extract metals, minerals, and coal)
> - Autonomous (fully robotic) mining
> - Geophysical techniques that can "see" through solid rock
> - Total resource recovery and/or waste utilization
> - Wasteless mining technologies
> - Minimal adverse environmental impacts
> - Rapid development of soils on mine wastes

Finding. The market will not support an optimal amount of research and development, possibly by a wide margin, without government support. The private sector tends to underfund research and development, particularly high-risk projects and projects with long-term benefits.

Finding. Although research in a broad range of fields may eventually have beneficial effects for the mining industry, the committee identified a number of areas in which new basic scientific data or technology would be particularly beneficial (Table 8-1).

Recommendation. The federal government has an appropriate, clear, and necessary role to play in funding research and development on mining technologies. The government should have a particularly strong interest in what is sometimes referred to as high-risk, "far-out," "off-the-path," or "blue-sky" research. A portion of the federal funding for basic research and long-term development should be devoted to achieving revolutionary advances with potential to provide substantial benefits to both the mining industry and the public.

Federal funding may be directed to agencies responsible for basic and applied research, including the Departments of Energy, Agriculture, Defense, Commerce, Health and Human Services, Interior, and Transportation, and NSF, EPA, and NASA. In addition, the Mining Industries of the Future Program should allow for support of long-range, "frontier," or "blue sky" projects. A side benefit of funding for basic research and long-term technology development, particularly in cooperation with universities, is the training of scientists and engineers in the mining sector and in other technology-intensive sectors of the economy.

AVAILABLE RESEARCH AND TECHNOLOGY RESOURCES

For more than a century the federal government has been involved in research and development for our basic industries. In addition, a number of federal agencies are involved in science, engineering, and technology development that could be useful to the mining industry.

Finding. The committee recognizes that federal agencies undertake worthwhile research and development for their own purposes. Research and development that could benefit

> **SIDEBAR 8-2**
> **Basic and Applied Research and Development**
>
> The Committee on Criteria for Federal Support of Research and Development (NRC, 1995c) developed the following characteristics for federally funded research:
>
> - **Basic Research** – creates new knowledge; is generic, nonappropriable, and openly available; is often done with no specific application in mind; requires a long-term commitment.
> - **Applied Research** – uses research methods to address questions with a specific purpose; pays explicit attention to producing knowledge relevant to producing a technology or service; overlaps extensively with basic research; can be short-term or long-term.
> - **Fundamental Technology Development** – develops prototypes; uses research findings to develop practical applications; is of general interest to a sector or sectors, but full returns cannot be captured by any one company; is usually short-term, but can be long-term; is not developed for one identifiable commercial or military product; often makes use of new knowledge from basic or applied research.

the mining sector of the U.S. economy is being pursued by many federal agencies. The problem is not the lack of skilled researchers but the lack of direct focus on the problems of most interest to the mining industry. It would be helpful if progress in these programs were systematically communicated to all interested parties, including the mining sector.

Recommendation. Because it may be difficult for a single federal agency to coordinate the transfer of research results and technology to the mining sector, a coordinating body or bodies should be established to facilitate transfer of appropriate federally funded technology to the mining sector. The Office of Industrial Technologies has made some progress in this regard by organizing a meeting of the agencies involved in research that could benefit the mining industry.

Office of Industrial Technology Mining Industries of the Future Program

The OIT is utilizing a consortia approach in its Industries of the Future Program. This model has proved extremely successful (NRC, 1997a). In programs focused on technologies for the mining industries such consortia should include universities, suppliers, national laboratories, and any ad hoc groups deemed to be helpful, as well as government entities and the mining industry.

The Mining Industries of the Future Program is subject to management and oversight by the Department of Energy and receives guidance from the National Mining Association and its Technology Committee. The committee recognizes that the research and technology needs of the mining industries draw upon many disciplines, ranging from basic sciences to applied health, safety, and environmental concerns.

Recommendation. Consortia are a preferred way of leveraging expertise and technical inputs to the mining sector, and the consortia approach should be continued wherever appropriate. Advice from experts in diverse fields would be helpful for directing federal investments in research and development for the mining sector. The Office of Industrial Technologies should institute periodic, independent program reviews of the Mining Industries of the Future Program to assure that industry needs are being addressed appropriately.

References

Abelson, P.H. 2000. Decreasing reliability of energy. Science 290(5493): 931.

Adamczyk, M. 2000. The next generation of longwall shearers. Coal Age 205 (9):74–78.

Alpers C. N., and D.W. Blowes, eds. 1994. Environmental Geochemistry of Sulfide Oxidation. Washington, D.C.: American Chemical Society Symposium.

American Geological Institute. 1997. Dictionary of Mining, Mineral, and Related Terms, 2nd ed. Alexandria, Va.: American Geological Institute.

Amils, R., and A. Ballester, eds. 1999a. Biohydrometallurgy and the Environment Toward the Mining of the 21st Century. Part A. Bioleaching Microbiology. Proceedings of the International Biohydrometallurgy Symposium, 1999. Amsterdam: Elsevier.

Amils, R., and A. Ballester, eds. 1999b. Biohydrometallurgy and the Environment Toward the Mining of the 21st Century. Part B. Molecular Biology, Biosorption, Bioremediation. Proceedings of the International Biohydrometallurgy Symposium, 1999. Amsterdam: Elsevier.

Australian Mineral Foundation. 1999. BIOMINE '99. Glenside, South Australia: Australian Mineral Foundation.

Avasthi, J.M., and A.H. Singleton. 1983. Underground coal gassification: a near-term alternate fuel from unmineable coal resources. Preprint 83-304. Reston, Va.: American Society of Civil Engineers.

Barnett, H.J., and C. Morse. 1963. Scarcity and Growth: The Economics of Natural Resource Availability. Resources for the Future. Baltimore, Md.: Johns Hopkins Press.

Bartlett, R.W. 1992. Solution Mining: Leaching and Fluid Recovery of Materials. Amsterdam: Gordon and Breach Science Publishers.

Bartlett, R.W. 1998. Solution Mining: Leaching and Fluid Recovery of Materials, 2nd Ed. Amsterdam: Gordon and Breach Science Publishers.

Bateman, A.M. 1942. The ore deposits of Kennicott, Alaska. Pp. 188–193 in Ore Deposits as Related to Structural Features, edited by W.H. Newhouse. Princeton, N.J.: Princeton University Press.

Bates, R.L., and J.A Jackson, eds. 1987. Glossary of Geology, 3rd ed. Alexandria: Va.: American Geological Institute.

Blowes, D.W., C.J. Ptacek, K.R. Waybrant, J.D. Bain, and W.D. Robertson. 1995. Treatment of mine drainage water using in situ permeable reactive walls. Pp. 979–987 in Proceedings of the Sudbury '95 - Mining and the Environment, Vol II. Sudsbury, Ontario: CANMET.

Carter, R.A. 1999. Jumbo equipment: haul more, haul faster, haul longer. Engineering and Mining Journal 200(7):28–32.

Catcheside, D.G.A., and J.P. Ralph. 1997. Biological Processing of Coal and Carbonaceous Materials. Pp. 11–18 in Proceedings of the 9th International Conference on Coal Science, Vol. 1. Essen, Germany: P&W Druck und Verlag GMBH.

Chi, G., M.C. Fuerstenau, R.C. Bradt, and A. Ghosh. 1996. Improved comminution efficiency through controlled blasting during mining. International Journal of Mineral Processes 47:93–101.

Chiaro, P.S., and G.F. Joklik. 1998. The extractive industries. Pp. 13–26 in The Ecology of Industry: Sectors and Linkages, edited by D.J. Richards and G. Pearson. Washington, D.C.: National Academy of Engineering.

Conroy, P.J., J.M. Heimerl, and E. Fisher. 2000. Rapid Excavation and Mining (REAM) System Revisited. Report ARL-TR-2249. Aberdeen Proving Ground, Md.: Army Research Laboratory.

Cox, D.P., and D.A. Singer, eds. 1992. Mineral Deposit Models, U.S.G.S. Bulletin. Washington, D.C.: U.S. Geological Survey.

Coyne, K.R., and Hiskey, J.B., eds. 1989. In-Situ Recovery of Minerals. New York: United Engineering Foundation.

Crozier, R.D. 1992. Oxide and non-metallic mineral flotation. Pp, 212–259 in Flotation Theory, Reagents, and Ore Testing, Oxford, U.K.: Pergamon Press.

Das, G.K., S. Acharya, S. Anand, and R.P. Das. 1996. Jarosite: a review. Mineral Processing and Extractive Metallurgy Review 6: 185–210.

Doyle, F.M., and A.H. Mirza. 1990. Understanding the mechanisms and kinetics of acid and heavy metals release from pyretic wastes. Pp. 43-51 in Mining and Mineral Processing Wastes, edited by F. Doyle. Littleton, Colo.: Society for Mining, Metallurgy, and Exploration.

EIA (Energy Information Administration). 1999a. U.S. Coal Reserves: 1997 Update. Washington, D.C. U.S. Department of Energy. Available online: http://www.eia.doe.gov/cneaf/coal/reserves/front-1.html

EIA. 1999b. Annual Energy Review. Washington, D.C.: U.S. Department of Energy. Available online: http://www.eia.doe.gov/emeu/aer/overview.html

EIA. 1999c. U.S. Coal Supply and Demand: 1999 Review. Washington, D.C.: U.S. Department of Energy.

EIA. 1999d. Uranium Industry Annual. Washington, D.C.: Energy Information Administration. Available online at: http://www.eia.doe.gov/cneaf/nuclear/uia/uia.pdf

EIA. 2001. World Coal Consumption, 1980-1999. Washington, D.C.: U.S. Department of Energy, Available online: http://www.eia.doe.gov/emeu/iea/Table14.html.

Eppinger, R.G., P.H. Briggs, D. Rosenkrans, and V. Ballestrazze. 2000. Environmental geochemical studies of selected mineral deposits in Wrangell-St. Elias National Park and Preserve, Alaska. U.S.Geological Survey Professional Paper 1619. Washington, D.C.: U.S. Geological Survey.

Fiscor, S. 1999. U.S. Longwall Census 1999. Coal Age 104(2): 30–35.

Fuerstenau, M.C., J.D. Miller, and M.C. Kuhn. 1985. Chemistry of Flotation. New York: Society of Mining Engineers.

Gao, M.W., K.R. Weller, and K.S.E. Forssberg. 1995. Energy particle size relationships in a stirred ball mill. Pp. 193–204 in Mineral Processing, edited by S.P. Mehrotra and R. Shekhar. New Delhi: Allied Publishers, Ltd.

REFERENCES

Gillette, R.S., R.A. Jankoswski, and F.N. Kissell. 1988. Increasing coal output will require better dust control. Pp. 52–55 in Proceedings of the 7th International Pneumoconiosis Conference. Pittsburgh, Pa.: U.S. Department of Health and Human Services.

Gobla, M., S. Schurman, and A. Sogue. 2000. Using Envirobond ARD to prevent acid rock drainage. Pp. 429-436 in Proceedings of the 7th International Conference on Tailings and Mine Waste '00. Rotterdam: A.A. Balkema.

Hartman, H.L. 1987. Introductory Mining Engineering. New York: John Wiley and Sons.

Hustrulid, W. 1982. Underground Metal Mining Methods Handbook. Littleton, Colo.: Society for Mining, Metallurgy, and Exploration.

Ives, K.J., ed. 1984. The Scientific Basis of Flotation. Boston, Mass.: Martinus Nijhoff Publishers..

Johnson, N.W. 1998. Application of the ISAMILL (horizontal stirred mill) to the lead zinc concentrate at Mt. Isa Mine Ltd. The Mining Cycle, Victoria, Australia: The Australasian Institute of Mining and Metallurgy.

Katen, K.P. 1992. Health and safety standards. Pp. 162–173 in SME Mining Engineering Handbook, edited by H.L. Hartman. Littleton, Colo.: Society for Mining, Metallurgy, and Exploration.

Khalafalla, S.E., and G.W. Reimers. 1981. Beneficiation with Magnetic Fluids. Report of Investigation #8532. Washington, D.C.: U.S. Bureau of Mines.

Kleinmann, R.L.P. 1997. Mine drainage systems. Pp. 237 in Mining Environmental Handbook, edited by J.J. Marcus. London: Imperial College Press.

Kleinmann, R.L.P., and R.S. Hedin. 1993. Treatment of mine water using passive methods. Pollution Engineering 25: 20–22.

Krautkraemer, J. 1998. Nonrenewable resource scarcity. Journal of Economic Literature 36: 2065–2107.

Kruger, D. 1994. Identifying Opportunities for Methane Recovery at U.S. Coal Mines (draft), EPA-430-R-94-012. Washington, D.C.: U.S. Environmental Protection Agency.

Lawrence, R.W. 1996. The management of acid rock drainage in the mining industry. Invited paper at VIII Encuentro Sobre Procesamiento de Minerales, Instituto de Metalurgia de la Universidad Autonoma de San Luis Potosi, Mexico, August 6–9, 1996.

Lawrence, R.W. 1998. Biotechnology in the mining industry. Pp. 319-326 in Advances in Biotechnology and Bioprocess Engineering, edited by O.T. Ramirez. Amsterdam: Kluwer Academic Publishers.

Lefond, S.J. 1975. Industrial Mineral and Rocks, 4th ed. New York: American Institute of Mining, Metallurgical, and Petroleum Engineers.

Leja, J. 1982. Surface Chemistry of Froth Flotation. New York: Plenum Press.

Lin, C.L., and J.D. Miller. 1997. Direct three-dimensional liberation analysis by cone beam X-ray microtomography. Pp. 85–94 in Comminution Practice, edited by S.K. Kawatra. Littleton, Colo.: Society for Mining, Metallurgy and Exploration.

Marshall, G.P., J.S. Thompson, and R.E. Jenkins. 1998. New Technology for Prevention of Acid Rock Drainage. Golden Colo.: Randol International, Ltd..

Mason, P.G, and J.W. Gulyas. 1999. Pressure hydrometallurgy: no longer regarded with trepidation for the treatment of gold and base metal ores and concentrates. Pp. 585–616 in EPD Congress 1999, edited by B. Mishra. Warrendale, Pa.: The Minerals, Metals and Materials Society.

McIvor, R.E. 1997. High pressure grinding rolls: A review. Pp. 95–98 in Comminution Practices, edited by S.K. Kowatra. Littleton, Colo.: Society for Mining, Metallurgy and Exploration.

Mitchell, P., C. Potter, and M. Watkins. 2000. Treatment of acid rock drainage: field demonstration of silica micro encapsulation technology and comparison with an existing caustic soda-based system. Pp. 312–313 in Proceedings of the 7th International Conference on Tailings and Mine Waste. Rotterdam: A.A. Balkema.

NIOSH (National Institute for Occupational Safety and Health). 1999. Mining Accident Involving Programmable Electronics. MSHA Workshop on Programmable Electronic Mining Systems: An Introduction to Safety, J.J. Sammarco (editor), NIOSH *Pittsburgh Research Laboratory*, Pittsburgh, PA, August 1999.

NIOSH. 2000. Health, Injuries, Illnesses, and Hazard Exposures in the Mining Industry, 1986–1995: A Surveillance Report. Washington D.C.: Centers for Disease Control and Prevention, U.S. Department of Health and Human Services.

National Safety Council. 1999. Injury Facts, 1999 Edition. Itasca, Ill.: National Safety Council.

NMA (National Mining Association). 1998a. The Future Begins with Mining. Washington, D.C.: National Mining Association.

NMA. 1998b. Mining Industry Roadmap for Crosscutting Technologies. Washington, D.C.: National Mining Association.

NMA. 1999. Facts about Coal, 1999–2000. Washington, D.C.: National Mining Association.

NMA. 2000. Mineral Processing Technology Roadmap. Washington, D.C.: Mining Industry of the Future. Available online: *http://www.oit.doe.gov/mining/mptroadmap.html*

NRC (National Research Council). 1990. Competitiveness of the U.S. Minerals and Metals Industry. Committee on Competitiveness of the U.S. Minerals and Metals Industry. Washington, D.C.: National Academy Press.

NRC. 1994a. Research Programs of the U.S. Bureau of Mines: First Assessment. Board on Earth Sciences and Resources. Washington, D.C.: National Academy Press. Available online: *http://books.nap.edu/catalog/9206.html*

NRC. 1994b. Drilling and Excavation Technologies for the Future. Committee on Advanced Drilling Technologies. Washington, D.C.: National Academy Press.

NRC. 1995a. Research Programs of the U.S. Bureau of Mines: 1995 Assessment. Board on Earth Sciences and Resources. Washington, D.C.: National Academy Press.

NRC. 1995b. Probabilistic Methods in Geotechnical Engineering. Board on Energy and Environmental Systems. Washington, D.C.: National Academy Press. Available online: *http://books.nap.edu/catalog/9476.html*

NRC. 1995c. Allocating Federal Funds for Science and Technology. Committee on Criteria for Federal Support of Research and Development. Washington, D.C.: National Academy Press. Available online: *http://books.nap.edu/catalog/5040.html*

NRC. 1996a. Mineral Resources and Society: A Review of the U.S. Geological Survey's Mineral Resource Surveys Program Plan. Board on Earth Sciences and Resources. Washington, D.C.: National Academy Press. Available online: *http://books.nap.edu/catalog/9035.html*

NRC. 1996b. Rock Fractures and Fluid Flow: Contemporary Understanding and Applications. Board on Energy and Environmental Systems. Washington, D.C.: National Academy Press. Available online: *http://books.nap.edu/catalog/2309.html*

NRC. 1997a. Intermetallic Alloy Development: A Program Evaluation. National Materials Advisory Board. Washington, D.C.: National Academy Press. Available online: *http://books.nap.edu/catalog/5701.html*

NRC. 1997b. Satellite Gravity and the Geosphere.: Contributions to the Study of the Solid Earth and Its Fluid Envelopes. Board on Earth Science and Resources. Washington, D.C.: National Academy Press. Available online: *http://books.nap.edu/catalog/5767.html*

NRC. 1998a. Manufacturing Process Controls for the Industries of the Future. Board on Manufacturing and Engineering Design. Washington, D.C.: National Academy Press. Available online: *http://books.nap.edu/catalog/6258.html*

NRC. 1998b. Seismic Signals from Mining Operations and the Comprehensive Test Ban Treaty: Comments on a Draft Report by a DOE Working Group. Board on Earth Sciences and Resources. Washington, D.C.: National Academy Press. Available online: *http://books.nap.edu/catalog/6226.html*

NRC. 1999a. Separation Technologies for the Industries of the Future. National Materials Advisory Board. Washington, D.C.: National Academy Press. Available online: *http://books.nap.edu/catalog/6388.html*

NRC. 1999b. Industrial Technology Assessments: An Evaluation of the Research Program of the Office of Industrial Technologies. National Materials Advisory Board. Washington, D.C.: National Academy Press. Available online: *http://books.nap.edu/catalog/9657.html*

NRC. 1999c. Hardrock Mining on Federal Lands. Board on Earth Sciences and Resources. Washington, D.C.: National Academy Press. Available online: *http://books.nap.edu/catalog/9682.html*

NRC. 2000. Seeing into the Earth: Noninvasive Characterization of the Shallow Subsurface for Environmental and Engineering Applications. Board on Earth Sciences and Resources. Washington, D.C.: National Academy Press. Available online: *http://books.nap.edu/catalog/5786.html*

NRC. 2001. Basic Research Opportunities in Earth Science. Board on Earth Sciences and Resources. Washington, D.C.: National Academy Press.

OIT (Office of Industrial Technology). 2000. Capabilities Matrix Mining. Prepared by the OIT Laboratory Coordinating Council. Washington, D.C.: U.S. Department of Energy. Available online: *http://www.oit.doe.gov/LCC/mining_matrix.shtml*

Orr, C., ed. 1977. Filtration Principles and Practices, Part I. New York: Marcel Dekker.

Orr, C., ed. 1979. Filtration Principles and Practices, Part II. New York: Marcel Dekker.

Orumwense, O.A., and E. Forssberg. 1992. Superfine and ultrafine grinding: A literature survey. Mineral Processing and Extractive Metallurgical Review 2: 107–127.

Parekh, B.K., and J.D. Miller, eds. 1999. Advances in Flotation Technology. Littleton, Colo.: Society for Mining, Metallurgy, and Exploration.

Phelps, R.W. 2000. Moving a mountain a day: Grassburg grows six-fold. Engineering and Mining Journal 201(6): 22–28.

Ramani, R.V., and J.M. Mutmansky. 2000. Mine health and safety at the turn of the millennium. Mining Engineering 51(9): 25–30.

Rawlings, D.E., ed. 1997. Biomining: Theory, Microbes and Industrial Processes. Berlin: Springer-Verlag.

Ripley, E.A., R.E. Redmann, and A.A. Crowder. 1996. Environmental Effects of Mining. Delray Beach, Fla.: St. Lucie Press.

Sailor, W.C., D. Bodansky, C. Braun, S. Fetter, and B. van der Zwaan. 2000. Nuclear power: A nuclear solution to climate change? Science 19(288): 1177–1178.

Schlitt, W.J., and J.B. Hiskey, eds. 1981. Interfacing technologies in solution mining. Pp. 370 in Proceedings of the 2nd SME-SPE International Solution Mining Symposium. New York: American Institute of Mining, Metallurgical, and Petroleum Engineers.

Schlitt, W.J., and D.A. Shock, eds. 1979. In situ uranium mining and ground water restoration. Pp. 137 in Proceedings of AIME Annual Meeting. New Orleans, La.: Society of Mining Engineers.

Shuey, S.A. 1999. Mining technology for the 21st century: Inco digs deeper. Engineering and Mining Journal 200(4): 18–24.

Simmons, G.L., J.N. Orlich, L.C. Lenz, and J.A. Cole 1999. Implementation and Start-up of N2TEC Flotation at the Lone Tree Mine. Presented at the Society of Mining Engineers Annual Meeting, Denver, Colorado, March 1–4, 1999.

Somasundarum, P., and B.M. Moudgil. 1987. Reagents in Mineral Technology. New York: Marcel Dekker, Inc.

Sparrow, G., and J.T. Woodcock. 1995. Cyanide and other lixiviant leaching systems for gold with some practical applications. Mineral Processing and Extractive Metallurgy Review 14: 193–247.

Staub, W.P., N.E. Hinkle, R.E. Williams, F. Anastasi, J. Osiensky, and D. Rogness. 1986. An Analysis of Excursions at Selected In Situ Uranium Mines in Wyoming and Texas. ORNL/TM-9956 and NUREG/CR-3967. Oakridge, Tenn.: Oak Ridge National Laboratory

Steffan, Robertson, and Kirsten, Inc. 1989. Draft Acid Rock Drainage, Technical Guide. Vol 1. Vancouver, B.C.: Bi Tech Publishers, Ltd.

Svarovsky, L., ed. 1977. Solid-Liquid Separation. Boston, Mass.: Butterworths.

Tippin R.B., H.L. Huiatt, and D. Butts. 1999. Silicate mineral and potash flotation. Pp. 199–212 in Advances in Flotation Technology B. K. Parekh and J.D. Miller, eds. Littleton, Colo.: Society for Mining, Metallurgy, and Exploration.

University of California. 1988. Mining Waste Study Final Report. Prepared by the Mining Waste Study Team of the University of Calfornia at Berkeley. Berkely, Calif.: University of California Press.

Uranium Institute. 1999. Statistics: Uranium Production Figures. Core Issues, Vol 2. Knightsbridge, London: Uranium Institute. Available online: *http://www.uilondon.org/coreissues/stats/uprod.htm*

U.S. Department of Labor. 1999. Mine Health and Safety Administration, Health Standards for Occupational Noise Exposure. Final Rule. Federal Register 64(176): 49565–49568.

U.S. Department of Labor. 2000a. The employment situation. News Release, October 6, 2000. Washington D.C.: Bureau of Labor Statistics.

U.S. Department of Labor. 2000b. Labor Day 2000 Statement by Davitt McAteer, Assistant Secretary of Labor, September 1, 2000. Washington, D.C.: Mine Safety and Health Administration.

USGS (U.S. Geological Survey). 2000. Mineral Commodity Summaries 2000. Washington, D.C.: U.S. Department of the Interior. Available online at: *http://minerals.usgs.gov/minerals/pubs/mcs/2000/mcs2000.pdf*

Wadsworth, M.E. 1983. Metallurgy: Past, present and future. Pp. 3–38 in Proceedings of the 3rd International Symposium on Hydrometallugy, edited by K. Osse and J.D. Miller. New York: American Institute of Mining, Metallurgical, and Petroleum Engineers.

Wheeler, P., and N. Walls. 1998. Surface Mining. Mining Annual Review, 1998, London: Mining Journal, Ltd.

Appendixes

A
Biographies of Committee Members

Milton H. Ward (*Chair*), president of Ward Resources, Inc., was previously president and chief executive officer of Cyprus Amax Minerals Company, president and chief operating officer of Freeport Minerals Company, and an officer of a number of other public companies. He was elected to the National Academy of Engineering for his leadership in developing, building, and operating major mineral production facilities in remote and challenging environments. He is currently a member of the advisory board of the Geoscience and Environmental Center, Sandia National Laboratories. Dr. Ward is former chairman of the Board of Directors of the American Mining Congress (predecessor to the National Mining Association). He served on Tulane University's Board of Administrators and as the Advisory Committee chairman for Tulane University Bioenvironmental Research Center. He received a B.S. and M.S. in mining engineering from the University of Alabama, an M.B.A. from the University of New Mexico, a Ph.D. from the University of London Royal School of Mining, and an honorary Ph.D. from the Colorado School of Mines.

Jonathan G. Price (*Vice-chair*) is state geologist and director of the Nevada Bureau of Mines and Geology. He was president of the American Institute of Professional Geologists in 1997 and is president of the Association of American State Geologists for 2000–2001. His prior experience includes positions with the Anaconda Company; U.S. Steel Corporation; Bureau of Economic Geology, University of Texas at Austin; and the National Research Council (NRC). His research and publications address mineral resources, geology and geochemistry of ore deposits, igneous petrology, tectonics, geologic hazards, geologic mapping, environmental geochemistry, and solution mining. He was a member of the NRC Board on Earth Sciences and Resources panel that produced *Mineral Resources and Society: A Review of the U.S. Geological Survey's Mineral Resource Surveys Program Plan (1996)* and *Hardrock Mining on Federal Lands (1999)*. He earned his B.A. in geology and German from Lehigh University and his Ph.D. in geology from the University of California, Berkeley.

Robert Ray Beebe, a consultant based in Tucson, Arizona, is a retired executive of both Homestake Mining Company and Newmont Mining Corporation. He received his B.S. and M.S. in metallurgical engineering from Montana School of Mines and holds that institution's Silver and Gold Medals. Early in his career, he taught at several mining schools and did research at the Mines Experiment Station of the University of Minnesota and Battelle Memorial Institute. He is a distinguished member of the Society of Mining Engineers, a member of the Minerals, Metals and Materials Society, a member and past president of the Mining and Metallurgical Society of America, and a member of the National Academy of Engineering. His areas of expertise include mining and mineral processing of ferrous and nonferrous metals. He has chaired the Bureau of Mines Advisory Board, the Mineral Engineering Advisory Committee of the University of California, Berkeley, and the Mineral Engineering Advisory Committee of Montana Tech. He is also a director at the National Advanced Drilling and Excavation Technologies Institute. A long-time NRC volunteer, Mr. Beebe has served on the National Materials Advisory Board and a number of committees, most recently as joint chair of the Committee on the Impact of Selling the National Helium Reserve.

Corale L. Brierley, an independent consultant, was chief of environmental process development at Newmont Mining Corporation, president of Advanced Mineral Technologies Inc., and chemical microbiologist, New Mexico Bureau of Mines and Mineral Resources. Her research interests include biogenic extractive metallurgy, biological treatment methods for inorganic wastes, and thermophilic chemautotrophic microorganisms. She is the author of 70 publications and holds 5 patents in the field of biotechnology applications in mineral processing and waste treatment. She served on the NRC Committee on Research Programs of the U.S. Bureau of Mines and the Committee on Ground Water Recharge. She was elected to the National Academy of Engineering for "innovations in applying biotechnology to mine production and remediation." Dr. Brierley obtained a B.S. and M.S. in biology and chemistry, respectively, from New Mexico Institute of Mining and Technology and a Ph.D. in environmental sciences from the University of Texas at Dallas.

Larry Costin, who has been with Sandia National Laboratories since 1978, is manager of the Geomechanics Department. Major areas of research and development in which he has been involved include static and dynamic fracture and fragmentation of brittle rock, constitutive modeling of brittle damaging materials, localization and shear banding in metals under dynamic loading, finite element modeling of rock structures using advanced constitutive models, design and fielding of large-scale in-situ geotechnical tests, and design and analysis of nuclear waste repository systems. He also has experience in management of large projects, application of quality assurance (NQA-1) standards to laboratory and field testing, environmental remediation, and hazardous waste management. He earned his Ph.D. in solid mechanics from Brown University.

Thomas Falkie, now chairman of Berwind Natural Resources Corporation, was the director of the U.S. Bureau of Mines, U.S. Department of the Interior (1974–1977). From 1969 to 1974, he was the head and chairman of the Mineral Engineering Department at Pennsylvania State University. From 1961 to 1969 he was employed by the International Minerals and Chemicals Corporation. Dr. Falkie obtained his Ph.D. in mining engineering from Pennsylvania State University. He was on the Advisory Committee on Mining and Mineral Resources Research, U.S. Department of the Interior, is past president of the American Institute of Mining, Metallurgical and Petroleum Engineers, and past president of the Society for Mining, Metallurgy, and Exploration. He is a member of the board of the National Mining Association and chairman of the American Coal Foundation. Dr. Falkie is the author of more than 200 publications, including book chapters and handbooks. He has received numerous awards and is a member of the National Academy of Engineering.

Norman L. Greenwald, now president of Norm Greenwald Associates in Tucson, Arizona, has also held positions at Woodward-Clyde Consultants, Newmont Mining Corporation, and Magma Copper Company. He has extensive experience in all aspects of environmental regulations and compliance: air quality; surface and ground water quality; management of hazardous and solid waste; environmental auditing, compliance, and management planning; legislative and regulatory development processes at both the state and federal levels. He received his M.S. in soil-water chemistry from the University of Arizona.

Kenneth N. Han is Regents Distinguished Professor and Douglas W. Fuerstenau Professor in the Department of Materials and Metallurgical Engineering at the South Dakota School of Mines and Technology (SDSM&T). He obtained his B.S. and M.S. from Seoul National University and his Ph.D. in metallurgical engineering from the University of California, Berkeley. Upon finishing his doctorate, Dr. Han joined the Department of Chemical Engineering, Monash University, in Melbourne, Australia, as lecturer and senior lecturer. He began his career at SDSM&T in 1981, and has been head of the Department of Metallurgical Engineering and dean of the College of Materials Science and Engineering. His research interests include hydrometallurgy, interfacial phenomena, metallurgical kinetics, solution chemistry, fine particle recovery, and electrometallurgy. He has published more than 130 papers in international journals and presented more than 100 papers at international conferences. Dr. Han is the author of 9 monographs and holds 8 patents. He has received numerous awards from academic, and technical, and professional societies. A member of the National Academy of Engineering, Dr. Han is currently a member of the Committee on Engineering Education for the National Research Council.

Murray Hitzman has been with Colorado School of Mines since 1996. Prior to this, he spent 11 years in the minerals industry. In addition to discovering the carbonate-hosted Lisheen Zn-Pb-Ag deposit in Ireland, he worked on porphyry copper and other intrusive-related deposits, precious metal systems, volcanogenic massive sulfide deposits, sediment-hosted Zn-Pb and Cu deposits, and iron oxide Cu-U-Au-LREE deposits throughout the world. For two years, he worked in Washington, D.C., first in the U.S. Senate and later in the White House Office of Science and Technology Policy, on environmental and natural resource issues. He has received numerous awards and published approximately 70 papers. His current interests focus on deposit-scale and district-scale studies of metallic ore systems. Deposit-scale studies examine the genesis of ore deposits through detailed field work and careful laboratory research to characterize the geologic setting of deposits and determine alteration and mineralization events. District-scale investigations involve geologic mapping to determine the tectonic and structural factors important in localizing mineral deposits and to evaluate regional-scale fluid flow and the geochemical processes involved in mineral deposit formation. He received his Ph.D. degree in geology from Stanford University.

Glenn Miller is currently director of the Center for Environmental Sciences and Engineering, University of Nevada, Reno, and professor in the Department of Environmental and Resources Sciences. His areas of interest include the fate and transport of organic compounds in soils and the atmosphere, the closure of precious-metals mines, and the treatment of acid mine waste drainage. He has been active in several environmental organizations related to mining during the past 20 years. He received his Ph.D. in agricultural chemistry from the University of California at Davis.

Dr. Miller was a member of the NRC Committee on Risk Assessment of Methyl Bromide and a reviewer of *Hardrock Mining on Federal Lands* (1999).

Raja V. Ramani holds the Anne B. and George H. Deike, Jr., Chair in Mining Engineering at Pennsylvania State University. A graduate of the Indian School of Mines, he holds an M.S. and Ph.D. in mining engineering from Penn State, where he has been on the faculty since 1970. His research activities include flow mechanisms of air, gas, and dust in mining environs; innovative mining methods, and health, safety, productivity, and environmental issues in the mineral industry. He has published more than 200 research papers, contributed to 25 books, and edited the proceedings of 15 national and international symposiums. He has been a consultant to the United Nations and the World Bank and has received numerous awards from academia and technical and professional societies. He was the 1995 president of the Society for Mining, Metallurgy, and Exploration and served on the U.S. Department of Health and Human Service's Mine Health Research Advisory Committee (1991–1998). Dr. Ramani was the chair of the National Academy of Sciences (NAS) Committee on Post Disaster Survival and Rescue (1979–1981), and a member of the Health Research Panel of the NAS Committee on Research Programs of the U.S. Bureau of Mines (1994). He was a member of the U.S. Department of the Interior's Advisory Board to the director of U.S. Bureau of Mines (1995) and a member of the Secretary of Labor's Advisory Committee on the Elimination of Coal Worker's Pneumoconiosis (1995–1996).

John E. Tilton is the William J. Coulter Professor of Mineral Economics in the Division of Economics and Business at the Colorado School of Mines and a university fellow at Resources for the Future. He is a former director of the Division of Economics and Business and a past president of the Mineral Economics and Management Society. His teaching and research interests over the past 30 years have focused on economic and policy issues associated with the metal industries and markets. His recent research has focused on the environment and mining, material substitution, long-run trends in metal demand, the recycling of metals, the sources of productivity growth in mining, and changes in comparative advantage in metal trade. He worked for a year as an Economic Affairs Officer for the Mineral and Metals Branch of the United Nations Conference on Trade and Development in Switzerland and spent two years at the International Institute for Applied Systems Analysis in Austria directing a research program on mineral trade and markets. More recently, he has been a visiting fellow at Resources for the Future in Washington, D.C.; a senior Fulbright scholar at the Ecole Nationale Superieure des Mines in Paris; and a visiting scholar at the Centro de Mineria at the Pontificia Universidad Catolica de Chile in Santiago. Dr. Tilton also served on various NRC boards and committees, most recently on the Panel on Integrated Environmental and Economic Accounting.

Robert Bruce Tippin is the research director of the Minerals Research Laboratory and adjunct professor in the Department of Material Science and Engineering at North Carolina State University. He has more than 30 years of experience worldwide in the technical management of mineral-related activities, including applied research, conceptual development, design, engineering, and project management from construction through start-up. He also has five years of plant operating experience. He has been involved with processes for a variety of commodities, including salt, potash, gold, copper, uranium, clay, bauxite, numerous industrial minerals, aggregates, and recycled metal recovery. Dr. Tippin is the author of more than 50 technical publications and has served on committees for various professional organizations. He received his M.S. in mineral engineering from the University of Alabama and his Ph.D. in metallurgical engineering from the University of Minnesota.

Rong-Yu Wan is manager of metallurgical research at Newmont Mining Corporation and adjunct professor in the Metallurgical Engineering Department, College of Mines and Earth Sciences at the University of Utah. Prior to this, she was research professor in the Metallurgical Engineering Department at the University of Utah; supervisor and chief of the Extractive Metallurgical Division, Beijing General Research Institute of Mining and Metallurgy; and project manager for the Beijing Mineral Processing Research Institute. She earned her B.S. in chemical engineering from Chiao Tung University, China, and her Ph.D. in metallurgy and metallurgical engineering from the University of Utah. In addition to directing research and development projects, she has done both fundamental and applied research in mineral processing and chemical metallurgical processes and has developed numerous innovative technologies. She is a member of the National Academy of Engineering.

NRC Staff

Tamara L. Dickinson is a senior staff officer for the NRC Board on Earth Sciences and Resources. She has served as program director for the Petrology and Geochemistry Program in the Division of Earth Sciences at the National Science Foundation and as discipline scientist for the Planetary Materials and Geochemistry Program at NASA Headquarters. As a postdoctoral fellow at the NASA Johnson Space Center, she conducted experiments on the origin and evolution of lunar rocks and highly reduced igneous meteorites. She holds a Ph.D. and M.S. in geology from the University of New Mexico and a B.A. in geology from the University of Northern Iowa.

B

Presentations to the Committee

The following individuals made presentations to the Committee on Technologies for the Mining Industries:

Corby Anderson, Montana Tech, Butte
Robert Baird, Kennecott Energy, Gilette, Wyoming
Jim Bartis, RAND, Santa Monica, California
David Beerbower, Peabody Group, St. Louis, Missouri
Kenneth Bennett, Caterpillar, Inc., Peoria, Illinois
George Bockosh, National Institute of Occupational Safety and Health, Pittsburgh, Pennsylvania
H.L. Boling, Consultant, Pima, Arizona
Jerry Cape, Consulting Engineer, Bradenton, Florida
T.T. Chen, CANMET, Ottawa, Canada
Stewart Clayton, Office of Fossil Energy, Department of Energy, Germantown, Maryland
Steve Cone, Cone Geochemical, Inc., Lakewood, Colorado
Steve Cotten, Consolidated Coal Company, Rices Landing, Pennsylvania
Les Darling, Knight-Piesold, Denver, Colorado
Kyle Dotson, BHP, Houston, Texas
John Finger, Sandia National Laboratory, Albuquerque, New Mexico
Bill Ford, National Stone Association, Washington, D.C.
Fred Fox, Kennecott Minerals Company, Salt Lake City, Utah
James Gephardt, Process Engineering Resources, Inc., Salt Lake City, Utah
Alexander Goetz, University of Colorado at Boulder
Todd Harris, Kline and Company, Fairfield, New Jersey
Mark Hart, Newmont Mining Corporation, Englewood, Colorado
Robin Hickson, Kvaerner Metals, San Ramon, California
Brent Hiskey, University of Arizona, Tucson
Steve Hoffman, Environmental Protection Agency, Washington, D.C.
Kate Johnson, U.S. Geological Survey, Reston, Virginia
Kathy Karpan, Office of Surface Mining, Reclamation, and Enforcement, Department of the Interior, Washington, D.C.
Paul Korpi, Cleveland Cliffs-Empire Mining Partnership, Palmer, Michigan
Martin Kuhn, Minerals Advisory Group, Tucson, Arizona
Victor Labson, USGS, Denver, Colorado
William Lane, Doe Run Company, Viburnum, Missouri
Richard Lawson, National Mining Association, Washington, D.C.
Pete Luckie, Pennsylvania State University, University Park
Deepak Malhotra, Resource Development, Inc., Wheat Ridge, Colorado
Toni Marechaux, Office of Industrial Technologies, Department of Energy, Washington, D.C.
Bill Maurer, Maurer Engineering, Inc., Houston, Texas
John Murphy, University of Pittsburgh, Pennsylvania
Haydn Murray, Indiana University, Bloomington
Carl Peterson, Massachusetts Institute of Technology, Cambridge
D.J. Peterson, RAND, Santa Monica, California
Robert Pruett, Imerys Pigments and Additives Group, Sandersville, Georgia
Jean-Michel Rendu, Newmont Mining Corporation, Denver, Colorado
Frank Roberto, Idaho National Engineering and Environmental Laboratory (INEEL), Idaho Falls, Idaho
Jim Rouse, Montgomery Watson, Golden, Colorado
Eric Seedorff, Specialty Products Systems, Tucson, Arizona
Peter Smeallie, Institute for Advanced Drilling, Washington D.C.
Dennis Stover, Rio Algom Mining Corporation, Oklahoma City, Oklahoma
Stanley Suboleski, Virginia Polytechnic Institute, Blacksburg
James Taranik, University of Nevada, Reno
Harry Tuggle, United Steelworkers of America, Pittsburgh, Pennsylvania
Ronald Wiegel, University of Minnesota, Coleraine
Roe-Hoan Yoon, Virginia Polytechnic Institute, Blacksburg
Sharon Young, Versitech Inc., Tucson, Arizona
Dirk Van Zyl, University of Nevada, Reno

C

Agency Web Addresses

Department of Agriculture — *www.usda.gov*
   U.S. Forest Service — *www.fs.fed.us*
   Inventory of national forest lands — *www.fs.fed.us/oonf/minerals/mgsite.htm*

Department of Commerce – *www.doc.gov*
   National Institute of Standards and Technology — *www.nist.gov*
   Advanced Technology Program — *www.atp.nist.gov*

Department of Energy — *www.energy.gov*
   The Office of Industrial Technologies — *www.oit.doe.gov*
   Mining Industries of the Future Program — *www.oit.doe.gov/mining*
   Office of Power Technologies — *www.eren.doe.gov/power*
   Albany Research Center — *www.alrc.doe.gov*
   National Renewable Energy Laboratory — *www.nrel.gov/st-it.html*
   Idaho National Engineering and Environmental Laboratory — *www.inel.gov*
   Argonne National Laboratory — *www.anl.gov*
   Los Alamos National Laboratory — *www.lanl.gov*
   Sandia National Laboratories — *www.sandia.gov*
   Office of Environmental Management — *www.em.doe.gov*
   Office of Transportation Technologies — *www.ott.doe.gov*

Department of Defense — *http://www.defenselink.mil/*
   Army Corps of Engineers Waterways Experiment Station — *www.wes.army.mil*
   Defense Advanced Research Projects Agency — *www.darpa.mil*

Department of Health and Human Services — *www.os.dhhs.gov*
   National Institute for Occupational Safety and Health — *www.cdc.gov/niosh*

Department of the Interior — *www.doi.gov/indexj.html*
   Bureau of Land Management — *www.blm.gov*
   Office of Surface Mining Reclamation and Enforcement — *www.osmre.gov*
   National Park Service — *www.nps.gov*
   U.S. Geological Survey — *www.usgs.gov*
   Mineral Resources Program — *www.minerals.er.usgs.gov*
     acid mine drainage and the development of technologies to remediate historic sites — *www.minerals.usgs.gov*
   Bureau of Reclamation — *www.usbr.gov/main/*
   Bureau of Indian Affairs — *www.doi.gov/bureau-indian-affairs.html*
   Division of Energy and Mineral Resources — *snake1.cr.usgs.gov/demr/index.htm*

Department of Labor — *www.dol.gov*
    Mine Safety and Health Administration — *www.msha.gov*
    Pittsburgh Safety and Health Technology Center — *www.msha.gov/TECHSUPP/TECHSUP1.HTM*
    Approval and Certification Center — *www.msha.gov/TECHSUPP/ACC/ACCHOME.HTM*

Department of Transportation — *www.dot.gov*
    Federal Railway Administration — *www.fra.dot.gov*
    Federal Transit Administration — *www.fta.dot.gov*

Environmental Protection Agency — *www.epa.gov*
    Copper Range Company Solution Mining Project — *www.epa.gov/Region5/copper*
    Mine Waste Technology Program — *www.epa.gov/ORD/NRMRL/std/mtb/annual99.htm*

National Aeronautics and Space Administration — *www.nasa.gov*
    Jet Propulsion Laboratory — *www.jpl.nasa.gov*

National Science Foundation — *www.nsf.gov*
    Directorate for Engineering — *www.eng.nsf.gov*
        Division of Civil and Mechanical Systems (tunneling and excavation research) — *http://www.eng.nsf.gov/cms/*
    Engineering Research Center for Particle Size and Technology at the University of Florida — *www.erc.ufl.edu*
    Division of Earth Sciences — *www.nsf.gov/search97cgi/vtopic*

Non-federal Programs
    National Aggregate Association/National Stone Association — *www.nationalaggregates.org/naa2.htm*
    National Ready-Mixed Concrete Association — *http://www.nrmca.org*

Acronyms

AIRES	Airborne Infrared Echelle Spectrometer	NIOSH	National Institute for Occupational Health and Safety
ASTER	Advanced Spaceborne and Thermal Emission and Reflection	NIST	National Institute of Standards and Technology
AVIRIS	Airborne Visible/Infrared Imaging Spectrometer	NMA	National Mining Association
DARPA	Defense Advanced Research Projects Agency	NRC	National Research Council
DOD	Department of Defense	NSF	National Science Foundation
DOE	Department of Energy	OIT	Office of Industrial Technology
EIA	Energy Information Administration	SITE	Superfund Innovative Technology Evaluation Program
GDP	Gross Domestic Product		
GPS	Global Positioning System	SXEW	Solvent Extraction and Eletrowinning
IOF	Industries of the Future	USBM	U.S. Bureau of Mines
MSHA	Mine Safety and Health Administration	USGS	U.S. Geological Survey
NASA	National Aeronautics and Space Administration		